REGULARITY THEORY
AND STOCHASTIC FLOWS FOR
PARABOLIC SPDEs

STOCHASTICS MONOGRAPHS
Theory and Applications of Stochastic Processes
A series of books edited by Mark Davis, Imperial College, London, UK

Volume 1
Contiguity and the Statistical
Invariance Principle
*P.E. Greenwood and
A.N. Shiryayev*

Volume 2
Malliavin Calculus for
Processes with Jumps
*K. Bichteler, J.B. Gravereaux
and J. Jacod*

Volume 3
White Noise Theory of
Prediction, Filtering and
Smoothing
*G. Kallianpur and
R.L. Karandikar*

Volume 4
Structure Selection
of Stochastic Dynamic Systems:
The Information Criterion
Approach
S.M. Veres

Volume 5
Applied Stochastic Analysis
*Edited by M.H.A. Davis and
R.J. Elliott*

Volume 6
Nonlinear Stochastic
Integrators, Equations and
Flows
R.A. Carmona and D. Nualart

Volume 7
Stochastic Processes
and Optimal Control
*Edited by H.J. Engelbert,
I. Karatzas and M. Röckner*

Volume 8
Stochastic Analysis and
Related Topics
*Edited by T. Lindstrøm,
B. Øksendal and A.S. Üstünel*

Volume 9
Regularity Theory and
Stochastic Flows for
Parabolic SPDEs
F. Flandoli

This book is part of a series. The publisher will accept continuation orders which may be cancelled at any time and which provide for automatic billing and shipping of each title in the series upon publication. Please write for details.

REGULARITY THEORY
AND STOCHASTIC FLOWS FOR
PARABOLIC SPDEs

by

Franco Flandoli
Scuola Normale Superiore
Pisa, Italy

CRC Press
Taylor & Francis Group
Boca Raton London New York

CRC Press is an imprint of the
Taylor & Francis Group, an **informa** business

First published 1995 by Gordon and Breach Science Publishers

Published 2021 by CRC Press
Taylor & Francis Group
6000 Broken Sound Parkway NW, Suite 300
Boca Raton, FL 33487-2742

ISBN 13: 978-2-88449-045-0 (hbk)

Visit the Taylor & Francis Web site at
http://www.taylorandfrancis.com

and the CRC Press Web site at
http://www.crcpress.com

British Library Cataloguing in Publication Data

Flandoli, Franco
  Regularity Theory and Stochastic Flows
  for Parabolic SPDEs. – (Stochastic Monographs,
  ISSN 0275-5785; Vol. 9)
  I. Title   II. Series
  515.353

Contents

Introduction to the Series   vii

Preface   ix

1   Introduction   1

2   Stochastic Flows: Preliminary Comments   2
   2.1   Regular Versions of Random Fields   3
   2.2   Comments on the Literature   5

3   Preliminaries on Well-Posedness and Function Spaces   8
   3.1   Notations   8
   3.2   Basic Abstract Well-Posedness Theorem   9
   3.3   Basic Particular Case   12
   3.4   An Extension   15
   3.5   Hilbert Scales and Interpolation of Random Field Spaces   17

4   Regularity in Bounded Domains: Some Counterexamples   18

5   Regularity Theory for Homogeneous Problems   24
   5.1   Abstract Regularity Theory   24
   5.2   Application to Equations in $\mathbf{R}^d$ 29
   5.3   Application to Second Order Equations in Bounded Domains: Dirichlet Boundary Condition   31
     5.3.1 $L^2(D)$-Solutions   32
     5.3.2 $H^1(D)$-Solutions   33
     5.3.3 $H^2(D)$-Solutions   38

5.3.4	$H^n(D)$-Solutions, $n \geq 3$	38
5.3.5	$H^{-1}(D)$-Solutions	39
5.3.6	$H^{-n}(D)$-Solutions, $n \geq 2$	39
5.3.7	Summary of Regularity Results for Dirichlet Boundary Value Problem. $H^\alpha(D)$-Solutions, $\alpha \in \mathbf{R}$	41
5.3.8	Regularity for the Adjoint Equation	41
5.4	Application to Equations of Order $2m$ in Bounded Domains: Dirichlet Boundary Conditions	42
5.5	Application to Second Order Equations in Bounded Domains: Neumann Boundary Condition	43
5.5.1	Regularity Results for the Neumann Boundary Value Problem	44
5.5.2	Regularity for the Adjoint Equation	47
6	**Non-Homogeneous Boundary Value Problems**	**48**
6.1	General Framework, Standing Assumptions, and Applications	48
6.2	Preliminaries on Deterministic Boundary Value Problems	51
6.3	Orientation	53
6.4	Elementary Regularity	53
6.5	More Refined Abstract Regularity Results	54
7	**Existence and Regularity of Stochastic Flows**	**57**
7.1	First Abstract Theorems of Existence and Regularity	57
7.2	A Pathwise Green Formula	59
7.3	Transpositions of the Adjoint Flow	62
7.4	Applications	66
7.4.1	Dirichlet Boundary Value Problem	66
7.4.2	Equations of Order $2m$ with Dirichlet Boundary Conditions	67
7.4.3	Neumann Boundary Condition	67
8	**An Alternative Approach**	**68**
Acknowledgements		76
References		76
Index		79

Introduction to the Series

The journal *Stochastics and Stochastics Reports* publishes research papers dealing with stochastic processes and their applications in modelling, analysis and optimization of systems subject to random disturbances. Stochastic models are now widely used in engineering, the physical and life sciences, economics, operations research and elsewhere. Moreover, these models are becoming increasingly sophisticated and often stretch the boundaries of a theory as it exists. A primary aim of *Stochastics and Stochastics Reports* is to further the development of the field by promoting an awareness of the latest theoretical developments on the one hand, and of all problems arising in applications on the other.

In association with *Stochastics and Stochastics Reports*, we are now publishing *Stochastics Monographs*, a series of independently produced volumes with the same aims and scope as the journal. *Stochastics Monographs* will provide timely and authoritative coverage of areas of current research in a more extended and expository form than is possible within the confines of a journal article. The series will include extended research reports, material derived from lecture courses on advanced topics, and multi-author works with a unified theme based on conference or workshop presentations.

Mark Davis

Preface

This volume contains treatment of two main topics in the theory of stochastic partial differential equations: space-regularity of solutions, and the existence of stochastic flows. In the former section, regularity theory in Sobolev space is covered comprehensively, for both homogeneous and non-homogeneous boundary value problems. This includes detailed analysis of the new geometrical conditions on coefficients, arising as a consequence of the stochasticity. The latter section comprises the application of regularity theory, to prove the existence of stochastic flows. A variety of results on stochastic flows are obtained by this method, and these are used to illustrate several open problems, with the hope of stimulating further research in this subject.

Throughout the book, the equations considered are linear parabolic, with multiplicative noise, analogous to those arising in nonlinear filtering, or diffusion models in randomly moving media. This volume provides an account of regularity results that will be a useful reference for any researcher studying stochastic partial differential equations.

REGULARITY THEORY AND STOCHASTIC FLOWS FOR PARABOLIC SPDE'S

FRANCO FLANDOLI

*Scuola Normale Superiore, Piazza dei Cavalieri 7,
56100 Pisa, Italy*

Stochastic evolution equations of parabolic type are considered. General regularity results for both homogeneous and non-homogeneous boundary value problems are given, and some counterexamples are discussed. The regularity theory is then used to give abstract criteria for the existence of stochastic flows. Several applications to concrete problems are discussed in detail.

KEY WORDS: Stochastic flows, stochastic evolution equations, regularity theory.

1   Introduction

Regularity theory of solutions to stochastic differential equations and existence of stochastic flows are related subjects. The existence of stochastic flows is a particular kind of regularity property. Even for finite dimensional stochastic equations two main approaches to the existence of stochastic flows are based on a careful analysis of the regularity of solutions with respect to initial data (cf. for instance [22], [18]). Although in a different way, also in the infinite dimensional case certain regularity results are among the main tools for the analysis of stochastic flows.

The purpose of this note is to review and improve some recent results in these related fields. The theory is rather well developed and clear for ordinary equations with delays and for parabolic equations in the full space. On the contrary, we have a less satisfactory understanding of stochastic parabolic equations in bounded domains. The book is addressed to the latter class of problems, with only some introductory remarks concerning the former two classes (section 2).

Regularity and flows for stochastic parabolic equations in bounded domains have been treated in [11], [16], [4], [12], [20] (see also [5] and [9] for other existence and regularity results). A number of open questions arises from these papers. The aim of the present work is to clarify the status of the problem after the above mentioned researches, and to give an answer to some of the questions left opened. In particular,

(i)   counterexamples are given to an exceedingly optimistic translation of classical deterministic regularity results to the stochastic case;

(ii)   the regularity theory for homogeneous and non-homogeneous stochastic parabolic equations in bounded domains is improved in various directions;

(iii)   a method of transposition is developed to obtain existence and regularity results for stochastic flows.

1

We start with a review section (2) on the problem of existence of regular versions of infinite dimensional random fields, and existence of stochastic flows for abstract evolution equations. Several fact discussed in section 2 reveals the importance of a good regularity theory for stochastic parabolic equations in view of the construction of the corresponding flows.

The problem of regularity of solutions is then trated in the subsequent sections 3–6. After some necessary preliminars (section 3), we discuss point (i) above. The counterexamples of section 4 show that, without certain geometrical restrictions on the coefficients of the differential operators in the diffusion terms (terms absent in the deterministic case), no solution exists of the homogeneous boundary value problem having a certain degree of regularity. These counterexamples reveal unexpected behaviours of stochastic boundary value problems.

In sections 5 and 6 we extensively study the regularity of solutions to homogeneous (section 5) and non-homogeneous (section 6) boundary value problems. Abstract results are first given, and then applied to Dirichlet and Neumann second order parabolic equations in bounded domains, with some remarks on higher order problems, and equations in $\mathbf{R}^d$. The picture given here is more complete than that of the previous papers [16], [4], [12], but still a number of problems are open.

Finally, starting from the introductory results of section 2.1 and the theory of section 5, the existence and regularity of stochastic flows for abstract and concrete parabolic equations in bounded domains is discussed (section 7). The main novelty here is the method of transposition (point (iii) above). It is inspired to the classical method known in the regularity theory for deterministic equations (cf. [25]). Its extension to the stochastic case is not trivial, because the requirement of adaptivity of solutions does not allow us to perform the time inversion of classical transposition in an obvious way. A different approach to transposition seems to be possible due to the recent work [35].

The abstract approach presented here to the existence of stochastic flows applies to a wide class of examples (see also [4] for other applications not treated here, to systems of equations and equations with periodic boundary conditions). It provide us with flows in several Sobolev spaces, not only in L^2−spaces. Moreover, such flows have regularity properties (of Hilbert-Schmidt type) that are important in some applications like in the analysis of Lyapunov exponents (cf. [16], [12]). However, the geometrical conditions imposed by the abstract approach (due to the fact that it is based on regularity theory) are not optimal. For second order parabolic equations with Dirichlet or Neumann boundary conditions, the existence of stochastic flows in L^2−spaces can be proved by other methods without the restrictions of the abstract approach. The simplest of these concrete methods (and the only one at present for Neumann boundary conditions) is presented in section 7.5. The comparison of the results of these approaches show that the theory of stochastic flows for infinite dimensional equations requires further understanding and investigation.

2   Stochastic Flows: Preliminary Comments

In contrast to the well developed theory of stochastic flows for stochastic ordinary differential equations (cf. [22], [18]), the results of existence and regularity of flows for infinite dimensional stochastic systems are rather fragmentary and uncomplete. Very basic questions are unsolved, and a good understanding is laking at various levels. To discuss the problem in more precise terms, let us give some preliminary definitions.

2.1 Regular Versions of Random Fields

Let H be a real separable Hilbert space and $(\Omega, \mathcal{F}, P)$ a complete probability space. Let $u(t, s; u_0)$, $s \leq t$, $u_0 \in H$, denote the solution at time t of a certain stochastic equation in H (over $(\Omega, \mathcal{F}, P)$), with given initial value u_0 at time s. The problem of existence of the *stochastic flow* is the problem of the existence of a *regular version* of the mapping $u_0 \rightarrow u(t, s; u_0)$ for fixed $s \leq t$ (or, when possible, uniformly in s and t). This means the existence of a mapping $\omega \rightarrow \phi_{s,t}(\omega)$ from Ω to the space of continuous mappings in H such that

$$\phi_{s,t}(\omega)u_0 = u(t, s, u_0)(\omega) \quad P - a.s. \tag{2.1}$$

for all $u_0 \in H$. $\phi_{s,t}(\omega)$ is called the stochastic flow in H associated to the given equation.

Let us first discuss for a moment the preliminary question of existence of a regular version of a given infinite dimensional random field. Let H and Y be two real separable Hilbert spaces with norms $|.|_H$ and $|.|_Y$. Lemma 2.1 below holds also in Polish spaces, and Lemma 2.2 in separable Banach spaces, with the same proofs, but we do not stress this generality. Let $(\Omega, \mathcal{F}, P)$ be a complete probability space, as above. Finally, let $L^0(\Omega; Y)$ be the space of Y-valued random variables. We call a (not necessarily linear) mapping $\Phi : H \rightarrow L^0(\Omega; Y)$ an Y-valued *random field* with parameter space H. Moreover, we say that Φ has a *continuous version* if there exists a mapping $\omega \rightarrow \phi(\omega)$, from Ω to $C(H, Y)$, the space of continuous mappings from H to Y, such that

$$\phi(\omega)x = (\Phi x)(\omega) \quad P - a.s. \tag{2.2}$$

for all $x \in H$. Just to settle down in technical terms the question of existence of regular version, let us prove the following elementary and essentially known Lemmas.

Lemma 2.1 *Let $\Phi : H \rightarrow L^0(\Omega; Y)$ be a given random field. Assume that for each ball S in H there exist two random variables $c_S(\omega) \geq 0$ and $\alpha_S(\omega) > 0$ such that*

$$|(\Phi x)(\omega) - (\Phi y)(\omega)|_Y \leq c_S(\omega)|x - y|_H^{\alpha_S(\omega)} \quad P - a.s. \tag{2.3}$$

for all $x, y \in S$. Then Φ has a continuous version ϕ (satisfying (2.2)), such that for P-a.e. $\omega \in \Omega$, $\phi(\omega)$ is Holder continuous on the balls of H, with the Holder constants given by (2.3). Finally, this regular version $\phi(\omega)$ is unique up to modifications on sets of measure zero.

Proof — Let $\mathcal{D}$ be a dense countable subset of H. For all $\omega \in \Omega$, define the mapping $\phi(\omega) : \mathcal{D} \to Y$ by means of the identity (2.2)). Since $\mathcal{D}$ is countable, there exists a set Ω_0, with $P(\Omega_0) = 1$, such that for all balls S, and all $x, y \in \mathcal{D} \cap S$ and all $\omega \in \Omega_0$ the inequality (2.3) holds true. Therefore, for all $x, y \in \mathcal{D} \cap S$ and all $\omega \in \Omega_0$ we have

$$|\phi(\omega)x - \phi(\omega)y|_Y \leq c_S(\omega)|x - y|_H^{\alpha_S(\omega)}.$$

Hence, when $\omega \in \Omega_0$, $\phi(\omega)$ is locally uniformly continuous from $\mathcal{D}$ to Y. Since $\mathcal{D}$ is dense in H, we can uniquely extend $\phi(\omega)$ to a mapping locally uniformly continuous from H to Y. On $\Omega - \Omega_0$ we can define $\phi(\omega)$ arbitrarily, completing in this way the definition of the mapping $\omega \to \phi(\omega)$. This family satisfies (2.2) when $x \in \mathcal{D}$ by definition. In general, if $x \in H$, and $\{x_n\} \subset \mathcal{D}$ is a sequence converging to x, it follows from (2.3) that $\Phi x_n \to \Phi x \ P - a.s.$. Thus, by (2.2), we have $\phi(\omega)x_n \to (\Phi x)(\omega) \ P - a.s.$. Since $\phi(\omega)$ is locally uniformly continuous for all $\omega \in \Omega$, we finally obtain (2.2).

As to the uniqueness, let $\phi'(\omega)$ be another locally uniformly continuous version (i.e. satisfying (2.2)). Since $\mathcal{D}$ is countable, it follows from (2.2) that there exists a set Ω_1, with $P(\Omega_1) = 1$, such that $\phi'(\omega)x = \phi(\omega)x$ for all $x \in \mathcal{D}$ and all $\omega \in \Omega_1$. This identity and the fact that both $\phi'(\omega)$ and $\phi(\omega)$ are locally uniformly continuous imply that $\phi'(\omega) = \phi(\omega)$ for all $\omega \in \Omega_1$, as claimed. The proof is complete.

We say that a random field $\Phi : H \to L^0(\Omega; Y)$ is *linear* if $\Phi(\lambda x + \mu y) = \lambda \Phi x + \mu \Phi y \ P - a.s.$, for all $x, y \in H$ and all $\lambda, \mu \in R$. In the case of linear random fields Lemma 2.1 gives the following result:

Lemma 2.2 *Let* $\Phi : H \to L^0(\Omega; Y)$ *be a given linear random field. Assume that there exists a random variable* $c(\omega) \geq 0$ *such that*

$$|(\Phi x)(\omega)|_Y \leq c(\omega)|x|_H \quad P - a.s. \tag{2.4}$$

Then Φ *has a continuous version* ϕ *such that* $\phi(\omega)$ *is a linear bounded operator from H to Y for a.e.* $\omega \in \Omega$. *The converse is also true.*

Proof — Let $\phi(\omega)$ be a version given by Lemma 2.1. We only have to prove that $\phi(\omega)$ is $P - a.s.$ a linear operator.

Since H is a separable Hilbert space, we can find a dense countable subset $\mathcal{D}$ of H which is also a vector space over Q. Since $\mathcal{D}$ and Q are countable, and Φ is linear, there exists a set Ω_0, with $P(\Omega_0) = 1$, such that

$$\begin{cases} \phi(\omega)x = (\Phi x)(\omega), \\ (\Phi(\lambda x + \mu y))(\omega) = \lambda(\Phi x)(\omega) + \mu(\Phi y)(\omega) \end{cases}$$

for all $x, y \in H$, $\lambda, \mu \in \mathbf{Q}$, $\omega \in \Omega_0$. Therefore, for all $\omega \in \Omega_0$, $\phi(\omega)$ is linear from $\mathcal{D}$ (as a vector space over $\mathbf{Q}$) to Y. By continuity, it is linear over H.

The converse statement is obvious.

Remark — In the finite dimensional case, a linear random field has always a regular version (for instance, it follows from Lemma 2.3 below). Thus the problem discussed here is typical of infinite dimensional random fields.

The case of *Hilbert-Schmidt* operators will be of particular interest. Next Lemma could be considered as a (very particular !) ∞-dimensional version of the classical regularity Theorem of Kolmogorov.

Lemma 2.3 *Let $\Phi : H \to L^2(\Omega; Y)$ be a linear random field. Assume that Φ is Hilbert-Schmidt from H to $L^2(\Omega; Y)$, i.e.*

$$\sum_{j=1}^{\infty} E|\Phi e_j|_Y^2 < \infty, \tag{2.5}$$

where $\{e_j\}$ is a complete orthonormal system in H. Then Φ has a regular version ϕ such that $\phi(\omega)$ is a Hilbert-Schmidt operator from H to Y for a.e. $\omega \in \Omega$.

Proof — For each $x \in H$ we have

$$|(\Phi x)(\omega)|_Y = |\sum_{j=1}^{\infty} < x, e_j >_H (\Phi e_j)(\omega)|_Y$$

$$\leq (\sum_{j=1}^{\infty} |(\Phi e_j)(\omega)|_Y^2)^{1/2}|x|_H \leq c(\omega)^{1/2}|x|_H \quad P - a.s. \tag{2.6}$$

where $c(\omega) = \sum_{j=1}^{\infty} |(\Phi e_j)(\omega)|_Y^2$. But $E[c] = \sum_{j=1}^{\infty} E|(\Phi e_j)|_Y^2 < \infty$, hence $c(\omega) < \infty$ $P - a.s.$. By Lemma 2.2, there exists a version ϕ of linear bounded operators from H to Y. Moreover, from (2.2) and the definition of $c(\omega)$ we have $c(\omega) = \sum_{j=1}^{\infty} |\phi(\omega)e_j|_Y^2$. Hence, since $c(\omega) < \infty$ $P - a.s.$, we conclude that $\phi(\omega)$ is Hilbert-Schmidt $P - a.s.$, as required.

A form of these and several other results can be found in [32]. An extension of Lemma 2.3 to the case of random fields of the form $u_0 \to u(t, s; u_0)$ (where the analysis of the dependence on t and s is essential) is proved in [12].

2.2 Comments on the Literature

After these preliminaries, we can continue the discussion of the present status of knowledge about infinite dimensional regular versions and flows. The problem of existence of stochastic flows is usually easy for equations with additive noise since they can be reduced to deterministic equations depending on a random parameter by a change of variable. Thus we shall concentrate on equations with non-constant diffusion coefficients.

A basic source of troubles is the *Wiener integral*. Taking $H = L^2(0, 1)$, $Y = R$, $w(t)$ a one dimensional Wiener process on $(\Omega, \mathcal{F}, P)$, the linear random field $\Phi : H \to L^2(\Omega; R)$ defined by the Wiener integral

$$\Phi f = \int_0^1 f(t) \, dw(t), \quad f \in H,$$ (2.7)

does not have a regular version, since it does not satisfy (2.4) ((2.4) would roughly correspond to have $\frac{dw(.,\omega)}{dt} \in L^2(0, 1)$ for a.e. $\omega \in \Omega$; for a rigorous proof see for instance [32]).

Based on this counterexample to the existence of regular versions, it is possible to construct counterexamples to the existence of stochastic flows for *stochastic delay equations*, [26]. The simplest problem is the delay equation

$$\begin{cases} dy(t) = y(t - 1) \, dw(t), \quad t \in [0, T] \\ y(0) = y_0 \in R; \quad y(.) = \psi(.) \in L^2(-1, 0) \quad \text{on} \quad [-1, 0]. \end{cases}$$ (2.8)

This equation does not generate a stochastic flow in $R \times L^2(-1, 0)$, since we have

$$y(1) = y_0 + \int_0^1 \psi(s - 1) \, dw(s),$$ (2.9)

and the Wiener integral does not have a linear continuous version from $L^2(0, 1)$ to R.

Equation (2.8) can be rewritten in abstract form in the Hilbert space $H = R \times L^2(-1, 0)$ with the classical device of [10]. Following [13], it takes the form

$$du(t) = Au(t) \, dt + Bu(t) \, dw(t), \quad u(0) = u_0,$$ (2.10)

where $u(t) = (y(t), y(.)|_{[t-1,t]}) \in H$, and the operators A and B (see the definitions in [13]) are linear unbounded operators in H. Thus (2.10) is an example of infinite dimensional stochastic equation which does not generates a stochastic flow.

Consider the more general *stochastic evolution equation* with multidimensional Brownian motion

$$\begin{cases} du(t) = Au(t) \, dt + \sum_{j=1}^{\infty} B^j u(t) \, dw^j(t), \\ u(0) = u_0. \end{cases}$$ (2.11)

The basic open question is to understand which abstract properties of A and B^j are responsable for the existence or non-existence of stochastic flows. For instance:

(i)   the answer is unknown even when A and B^j are bounded linear operators;

(ii) when A and B^j correspond to equation (2.10) the flow does not exist; when they correspond to certain parabolic equations the flow exists. In both cases A and B^j are unbounded operators. A surprising fact is that it is possible to unify delay and parabolic equations under a general abstract assumption ([14]).

The previous two facts lead to the conjecture that the regularization property typical of parabolic equations may be a basic tool for the construction of flows (in a sense, the case of bounded A and B^j is less regular than the parabolic case). Note that also Lemma 2.3 suggests to look for regularity properties of the random fields.

Related to parabolic problems are also the works [30], [22], [37], [16]. In [30], the solution $u(t, x, \omega)$ of a second order parabolic equation in R^d is represented by the formula

$$u(t, x, \omega) = \int_{R^d} K(t, x, y, \omega) u_0(y) \, dy \qquad (2.12)$$

where $K(t, x, y, \omega)$ is the so called *random fundamental solution*, i.e. it is solution in (t, x) of the stochastic equation, and satisfies the initial condition $K(0, x, y, \omega) = \delta(x - y)$. The authors prove the existence of $K(t, x, y, \omega)$ by means of regularity results. From (2.12) and the regularity of $K(t, x, y, \omega)$ it follows the existence of the stochastic flow. The same method has been used in [16], in the case of parabolic equations in bounded domains, in order to prove a compactness property of the flow. In bounded domains, formula (2.12) has been proved only under certain geometric restrictions on the coefficients of the parabolic equation.

This approach may be consedered as a concrete counterpart of the theory developed in sections 5 and 7 below. Of course, the advantage of the abstract theory is to cover a large number of different parabolic problems.

A second concrete method for stochastic parabolic equations in R^d is based on the so called *stochastic characteristics*, see [37]. This approach transforms the stochastic equation into a deterministic equation depending on the parameter $\omega \in \Omega$, and gives again a representation formula that can be used to prove existence and regularity of the stochastic flow.

A third approach is based on the representation of solutions by means of *Feynman-Kac type formulas* (related in a sense to (2.12)). This method reduces the problem of existence of the flow for the partial differential equation to the analysis of the flow associated to a finite dimensional stochastic equation. The Feynman-Kac approach has been used by [22] for parabolic equations in R^d, by [16] for parabolic equations in a bounded domain of R^d with Dirichlet boundary conditions, and by [31] for a special class of first order stochastic partial differential equations in bounded domains (transport equations).

As we remarked in the introduction, one of the main open questions is to understand why the Feynmann-Kac approach, or the easier approach presented in section 7.5, do not require the geometric restrictions on $b^j(x)$ imposed by the abstract approach of sections 5 and 7.

Finally, we mention the works [7], [8], [3], [32], where other results related to flows can be found. In particular, it is worth noting the *robust equation* approach, [7], [8], which gives, as a by-product of the existence of solutions, the existence of flows; the main restriction is that the B^j's must commute.

As to nonlinear flows, the author is only aware of the robust equation approach [8], [9], a trick based on the *skew-symmetry* of B^j [3], a generalization of the method of section 7.5 to reaction-diffusion equations [15], and the following natural idea. Consider a *semilinear equation* of the form

$$\begin{cases} du(t) = Au(t) \, dt + F(u(t)) \, dt + Bu(t) \, dw(t), & t \in [0, T] \\ u(0) = u_0 \in H. \end{cases} \qquad (2.13)$$

Assume that the linear equation with $F = 0$ generates a stochastic flow $\phi_{s,t}$. Then rewrite equation (2.13) in the integral form

$$u(t) = \phi_{0,t}u_0 + \int\limits_0^t \phi_{s,t} F(u(s)) \, ds. \tag{2.14}$$

A direct analysis of equation (2.14) should provide in some cases the existence of the nonlinear flow associated to (2.13). This method has not been developed in the literature yet. In connection with the flows constructed in section 7, it is important to note that $\phi_{s,t}$ is singular for $t = s$ in our case (at least a priori), so that we cannot treat equation (2.14). To study (2.14) one needs a better control for $\phi_{s,t}$ in s and near the diagonal, which more likely may follow from the explicit representations of Feynmann-Kac type, stochastic characteristics, of type (2.12), or as in section 7.5.

3 Preliminaries on Well-Posedness and Function Spaces

The framework and results developed in this section, based on a *scale of Hilbert spaces* and on *coercivity* assumptions in integral form like (3.5) (introduced in [14] with other aims), may appear too general in view of the application to second order parabolic problems with regular coefficients. Indeed, as far as the basic existence and uniqueness in $L^2(D)$ is concerned, the variational approach [27] (easier and more classical than the present one) is sufficient. However, for the analysis of general (not necessarily selfadjoint and reduced only to higher order terms) second order problems in suitable scales of Sobolev spaces the variational approach is less convenient or inapplicable.

3.1 Notations

Let $(\Omega, \mathcal{F}, \mathcal{F}_t, P)$ be a complete stochastic basis. All Brownian motions in the sequel will be defined on this basis. Let H be a Hilbert space. Given t, let $L^2(\mathcal{F}_t, H)$ be the space of square integrable random variables with values in H, measurable with respect to $\mathcal{F}_t$. Given $a < b$, let $C_{\mathcal{F}}(a, b; H)$ the space of continuous stochastic processes $u : [a, b] \times \Omega \to H$ which are adapted to the filtration $\mathcal{F}_t$, and such that

$$|u|^2_{C_{\mathcal{F}}(a,b;H)} := E \sup_{t \in [a,b]} |u(t)|^2_H < \infty.$$

Similarly, let $L^2_{\mathcal{F}}(a, b; H)$ the space of progressively measurable stochastic processes $u : [a, b] \times \Omega \to H$, such that

$$|u|^2_{L^2_{\mathcal{F}}(a,b;H)} := E \int\limits_a^b |u(t)|^2_H < \infty.$$

Since we shall mainly deal with a given time interval $[0, T]$, we shall denote $C_{\mathcal{F}}(0, T; H)$ and $L^2_{\mathcal{F}}(0, T; H)$ simply by $C_{\mathcal{F}}(H)$ and $L^2_{\mathcal{F}}(H)$.

3.2 Basic Abstract Well-Posedness Theorem

Let $\mathcal{Y} \subset \mathcal{X}$ be two real separable Hilbert spaces. We denote by $|.|_{\mathcal{Y}}$ and $< ., . >_{\mathcal{Y}}$ the norm and inner products in $\mathcal{Y}$, and we adopt analogous notations for $\mathcal{X}$. All inclusions in this section are understood in the set-theoretical and topological sense, while the density as to be specified explicitly.

On the stochastic basis $(\Omega, \mathcal{F}, \mathcal{F}_t, P)$, let $\{w^j(t) : t \geq 0, j \in \mathbf{N}\}$ be a sequence of independent 1-dimensional standard Wiener processes. We deal with the stochastic evolution equation in $\mathcal{X}$

$$u(t) = \phi(t) + \sum_{j=1}^{\infty} \int_0^t e^{(t-s)A} B^j u(s) \, dw^j(s) \tag{3.1}$$

where the assumptions on A and the sequence $(B) = \{B^j ; j \in \mathbf{N}\}$ will be given below.

Let us introduce the spaces and topologies related to this equation. In the applications the operators B^j will be unbounded in $\mathcal{X}$, with the property

$$B^j \in L(\mathcal{Y}, \mathcal{X}).$$

Hence we shall look for solutions $u \in L^2_{\mathcal{F}}(\mathcal{Y})$. However, a further condition on u arises from equation (3.1):

$$\sum_{j=1}^{\infty} E \int_0^T |B^j u(t)|^2_{\mathcal{X}} \, dt < \infty. \tag{3.2}$$

We denote by $\mathcal{Y}_{(B)}$ the space of all $y \in \mathcal{Y}$ such that

$$\sum_{j=1}^{\infty} |B^j y|^2_{\mathcal{X}} < \infty,$$

and by $L^2_{\mathcal{F}}(\mathcal{Y}_{(B)})$ the space of all $u \in L^2_{\mathcal{F}}(\mathcal{Y})$ such that condition (3.2) is satisfied. In $L^2_{\mathcal{F}}(\mathcal{Y}_{(B)})$ we define the seminorm

$$[u]^2_{L^2_{\mathcal{F}}(\mathcal{Y}_{(B)})} = \sum_{j=1}^{\infty} E \int_0^T |B^j u(t)|^2_{\mathcal{X}} \, dt$$

and the norm

$$||u||^2_{L^2_{\mathcal{F}}(\mathcal{Y}_{(B)})} = |u|^2_{L^2_{\mathcal{F}}(\mathcal{Y})} + [u]^2_{L^2_{\mathcal{F}}(\mathcal{Y}_{(B)})}.$$

In general, $\mathcal{Y}_{(B)}$ (resp. $L^2_{\mathcal{F}}(\mathcal{Y}_{(B)})$) is strictly included in $\mathcal{Y}$ (resp. in $L^2_{\mathcal{F}}(\mathcal{Y})$), and $L^2_{\mathcal{F}}(\mathcal{Y}_{(B)})$ is not complete with respect to $\|.\|^2_{L^2_{\mathcal{F}}(\mathcal{Y}_{(B)})}$. When the operators B^j are preclosed, $L^2_{\mathcal{F}}(\mathcal{Y}_{(B)})$ is complete; this happens in the standard applications to SPDE's because the B^j's are differential operators, but we do not need to assume the preclosedness of B^j at the abstract level. Obviously, if

$$\sum_{j=1}^{\infty} |B^j x|^2_{\mathcal{X}} \le c|x|^2_{\mathcal{Y}} \quad \forall x \in \mathcal{Y}, \tag{3.3}$$

then $\mathcal{Y}_{(B)} = \mathcal{Y}$ and $L^2_{\mathcal{F}}(\mathcal{Y}_{(B)}) = L^2_{\mathcal{F}}(\mathcal{Y})$. This is the case, for instance, when only a finite number of the B^j's are different from zero, as in most of our application (but this restriction will be imposed mainly for other reasons). However, some results do not depend on such properties.

We first recall a regularity result in $L^2_{\mathcal{F}}(\mathcal{Y})$ from [14], which is proved by elementary stochastic calculus (see also step 1 of the next Theorem).

Lemma 3.1 *Let $A : D(A) \subset \mathcal{X} \to \mathcal{X}$ be the infinitesimal generator of a strongly continuous semigroup e^{tA} in $\mathcal{X}$, such that $D(A) \subset \mathcal{Y}$, and*

$$\int_0^T |e^{tA}x|^2_{\mathcal{Y}} \, dt \le c_1 |x|^2_{\mathcal{X}}, \quad \forall x \in D(A).$$

Let $(f) = \{f^j \; ; \; j \in N\} \subset L^2_{\mathcal{F}}(\mathcal{X})$ be a sequence such that

$$\sum_{j=1}^{\infty} E \int_0^T |f^j(t)|^2_{\mathcal{X}} \, dt < \infty.$$

Then the series

$$I(f)(t) = \sum_{j=1}^{\infty} \int_0^t e^{(t-s)A} f^j(s) \, dw^j(s)$$

converges in $L^2_{\mathcal{F}}(\mathcal{Y})$, and

$$|I(f)|^2_{L^2_{\mathcal{F}}(\mathcal{Y})} \le c_1 \sum_{j=1}^{\infty} E \int_0^T |f^j(t)|^2_{\mathcal{X}} \, dt.$$

We can now state the main abstract result of this section. This Theorem has already been proved in [14], but we give here a new proof which readily extends to the more general case treated in section 3.4 (which applies to non-homogeneous boundary value problems). Theorem 3.2 is stated separately from Theorem 3.5 of that section for simplicity of exposition and later reference.

Theorem 3.2 *Let* $\mathcal{Y} \subset \mathcal{X}$ *and* $A : D(A) \subset \mathcal{X} \to \mathcal{X}$ *be the infinitesimal generator of a strongly continuous semigroup* e^{tA} *in* $\mathcal{X}$, *such that* $D(A) \subset \mathcal{Y}$, *and*

$$\int_0^T |e^{tA}x|_{\mathcal{Y}}^2 \, dt \leq c_1 |x|_{\mathcal{X}}^2, \quad \forall x \in D(A). \tag{3.4}$$

Let $B^j \in L(\mathcal{Y}, \mathcal{X})$, $j \in N$, *be linear operators satisfying*

$$\sum_{j=1}^\infty \int_0^\tau |B^j e^{tA} x|_{\mathcal{X}}^2 \, dt \leq (\eta + c_2 \tau)|x|_{\mathcal{X}}^2, \tag{3.5}$$

for all $x \in D(A)$ *and* $\tau \in [0, T]$, *and for some constants* $\eta \in (0, 1)$ *and* $c_2 > 0$. *Finally, let* $\phi \in L_{\mathcal{F}}^2(\mathcal{Y}_{(B)})$ *be given.*
Then equation (3.1) has a unique solution u *in* $L_{\mathcal{F}}^2(\mathcal{Y}_{(B)})$, *and*

$$||u||_{L_{\mathcal{F}}^2(\mathcal{Y}_{(B)})}^2 \leq c ||\phi||_{L_{\mathcal{F}}^2(\mathcal{Y}_{(B)})}^2$$

for some constant $c > 0$ *independent of* ϕ.

Proof Step 1— *We establish a preliminary property. Fix* $\tau \in [0, T]$. *If* (f) *is a sequence as in Lemma 3.1 (with* τ *in place of* T), *then the series* $I(f)$ *of Lemma 3.1 defines an element of* $L_{\mathcal{F}}^2(0, \tau; \mathcal{Y})$. *In addition,* $I(f) \in L_{\mathcal{F}}^2(0, \tau; \mathcal{Y}_{(B)})$ *and*

$$[I(f)]_{L_{\mathcal{F}}^2(0,\tau;\mathcal{Y}_{(B)})}^2 = \sum_{k=1}^\infty E \int_0^\tau |B^k \sum_{j=1}^\infty \int_0^t e^{(t-s)A} f^j(s) \, dw^j(s)|_{\mathcal{X}}^2 \, dt$$

$$\leq (\eta + c_2 \tau) \sum_{j=1}^\infty E \int_0^\tau |f^j(t)|_{\mathcal{X}}^2 \, dt. \tag{3.6}$$

The proof is similar to that of Lemma 3.1. It is elementary if $f^j \in L_{\mathcal{F}}^2(0, \tau; D(A))$ *and only a finite number of* f^j *are different from 0; the general case follows by approximation.*

Step 2 — Let us solve locally equation (3.1) by successive approximations. Let $u_0 = \phi$,

$$u_{n+1}(t) = \phi(t) + \sum_{j=1}^\infty \int_0^t e^{(t-s)A} B^j u_n(s) \, dw^j(s).$$

By the result of step 1 and the assumption on ϕ, we have by induction that $u_n \in L_{\mathcal{F}}^2(0, \tau; \mathcal{Y}_{(B)})$, and

$$[u_{n+1} - u_n]_{L_{\mathcal{F}}^2(0,\tau;\mathcal{Y}_{(B)})}^2 \leq (\eta + c_2 \tau)[u_n - u_{n-1}]_{L_{\mathcal{F}}^2(0,\tau;\mathcal{Y}_{(B)})}^2.$$

Hence

$$[u_{n+1} - u_n]^2_{L^2_{\mathcal{F}}(0,\tau;\mathcal{Y}_{(B)})} \leq (\eta + c_2\tau)^n [u_1 - \phi]^2_{L^2_{\mathcal{F}}(0,\tau;\mathcal{Y}_{(B)})}. \tag{3.7}$$

On the other side, by the estimate in Lemma 3.1,

$$|u_{n+2} - u_{n+1}|^2_{L^2_{\mathcal{F}}(0,\tau;\mathcal{Y})} \leq c_1 [u_{n+1} - u_n]^2_{L^2_{\mathcal{F}}(0,\tau;\mathcal{Y}_{(B)})}$$

$$\leq c_1(\eta + c_2\tau)^n [u_1 - \phi]^2_{L^2_{\mathcal{F}}(0,\tau;\mathcal{Y}_{(B)})}.$$

If τ is sufficiently small, we obtain that u_n converges in $L^2_{\mathcal{F}}(0, \tau; \mathcal{Y})$ to some process u. Using also (3.7), one can show that $u \in L^2_{\mathcal{F}}(0, \tau; \mathcal{Y}_{(B)})$, and can take the limit in equation (3.1) to prove that u satisfies (3.1). The uniqueness in $L^2_{\mathcal{F}}(0, \tau; \mathcal{Y}_{(B)})$ follows from equation (3.1), since if u_1 and u_2 are solutions, from (3.1) and (3.6) we have

$$[u_1 - u_2]^2_{L^2_{\mathcal{F}}(0,\tau;\mathcal{Y}_{(B)})} \leq (\eta + c_2\tau)[u_1 - u_2]^2_{L^2_{\mathcal{F}}(0,\tau;\mathcal{Y}_{(B)})},$$

whence $u_1 = u_2$.

Finally, the local procedure can be repeated over intervals of constant lenght, since the only condition on τ is $\eta + c_2\tau < 1$. The proof of the last inequality proceeds similarly an in standard way. The Theorem is proved.

3.3 Basic Particular Case

In view of our applications, let us discuss the assumptions of Theorem 3.2. It is known (cf. [14]) that an analytic semigroup of contractions satisfies the first assumption (3.4) with the choice $\mathcal{Y} = D((-A)^{\frac{1}{2}})$. For more general conditions see [14]. Moreover, a more algebraic condition implying (3.5) is

$$\frac{1}{2}\sum_{j=1}^{\infty} |B^j u|^2_{\mathcal{X}} \leq -\eta < Au, u >_{\mathcal{X}} +\lambda|u|^2_{\mathcal{X}}, \quad u \in D(A),$$

for some constants $\eta \in (0, 1)$ and $\lambda > 0$; see again [14] (the proof of this fact is strightforward). Finally, for sake of simplicity, we shall assume condition (3.3); this implies $L^2_{\mathcal{F}}(\mathcal{Y}_{(B)}) = L^2_{\mathcal{F}}(\mathcal{Y})$ and simplifies both the exposition and some technical aspects, as the interpolation results of section 5. As we said, (3.3) is satisfied in the applications that we have in mind. We arrive to the following corollary, directly used in section 5. Here the pathwise continuity is guaranteed by a classical result (cf. [19]), and is due to the fact that the semigroup is of contractions. Note that e^{tA} is a contraction semigroup (up to a translation of A) because of assumption (3.8).

Corollary 3.3. *Let $\mathcal{X}$ be a real separable Hilbert space, and let $A : D(A) \subset \mathcal{X} \to \mathcal{X}$ be the infinitesimal generator of an analytic semigroup of negative type and let*

$$\mathcal{Y} = D((-A)^{\frac{1}{2}}).$$

Assume that B^j, $j \in N$, be linear operators satisfying the following conditions:

$$B^j \in L(\mathcal{Y}, \mathcal{X});$$

$$\frac{1}{2} \sum_{j=1}^{\infty} |B^j u|_{\mathcal{X}}^2 \leq -\eta < Au, u >_{\mathcal{X}} + \lambda |u|_{\mathcal{X}}^2, \quad u \in D(A), \tag{3.8}$$

and

$$\sum_{j=1}^{\infty} |B^j u|_{\mathcal{X}}^2 \leq c_3 |u|_{\mathcal{Y}}^2, \quad u \in \mathcal{Y}, \tag{3.9}$$

for some constants $\eta \in (0, 1)$, $\lambda > 0$, $c_3 > 0$.

Then, for all $\phi \in L_{\mathcal{F}}^2(\mathcal{Y})$ there exists a unique solution u of equation (3.1) in $L_{\mathcal{F}}^2(\mathcal{Y})$. If in addition $\phi \in C_{\mathcal{F}}(\mathcal{X})$, then also $u \in C_{\mathcal{F}}(\mathcal{X})$, and

$$|u|_{C_{\mathcal{F}}(\mathcal{X})}^2 + |u|_{L_{\mathcal{F}}^2(\mathcal{Y})}^2 \leq c(|\phi|_{C_{\mathcal{F}}(\mathcal{X})}^2 + |\phi|_{L_{\mathcal{F}}^2(\mathcal{Y})}^2) \tag{3.10}$$

for some constant $c > 0$ independent of ϕ.

Finally, in particular, the function ϕ defined as

$$\phi(t) = e^{tA} u_0,$$

with $u_0 \in L^2(\mathcal{F}_0; \mathcal{X})$ belongs to $C_{\mathcal{F}}(\mathcal{X}) \cap L_{\mathcal{F}}^2(\mathcal{Y})$.

Proof — Recalling the previous remarks, we can apply Theorem 3.2 to obtain the existence and uniqueness in $L_{\mathcal{F}}^2(\mathcal{Y})$. If $\phi \in C_{\mathcal{F}}(\mathcal{X})$, from equation (3.1) and the result of [9], Theorem 6.10, we have that $u \in C_{\mathcal{F}}(\mathcal{X})$ and (3.10) holds true. In order to apply the result of [9], we represent the convolution integral of (3.1) in the form

$$\int_0^t e^{(t-s)A} \psi(u(s)) dw(s)$$

where $w(t) = \sum_{j=1}^{\infty} w^j(t) e_j$ is a cylindrical Wiener process in $\mathcal{X}$, $\{e_j\}$ is a complete orthonormal system in $\mathcal{X}$, and $\psi(y)$, $y \in \mathcal{Y}$, is defined as $\psi(y) = \sum_{j=1}^{\infty} B^j y \otimes e_j$. Notice that, by assumption (3.9), $u \in L_{\mathcal{F}}^2(\mathcal{Y})$ implies $\psi(u) \in L_{\mathcal{F}}^2(0, T; L_2(\mathcal{X}))$, where $L_2(\mathcal{X})$ is the space of Hilbert-Schmidt operators in $\mathcal{X}$, and $L_{\mathcal{F}}^2(0, T; L_2(\mathcal{X}))$ is defined similarly to $L_{\mathcal{F}}^2(0, T; \mathcal{X})$. Therefore, we can use the result of [9].

Finally, the last statement is due to the strong continuity of e^{tA} and property (3.4). The proof is complete.

Remark 1 — Assumption (3.8) is only used to have (3.5) and that the semigroup is of contractions. Thus the latter two assumptions may be substituted to (3.8), and the estimate (3.10) still holds true in such case. This fact will be used in the next proof.

Associated to Corollary 3.3 we prove an approximation result which will be used in the proof of Lemma 7.4.

Lemma 3.4 *Assume the same conditions of Corollary 3.3. Let $J_n = n(n - A)^{-1}$ be the approximations of the identity appearing in the Yosida approximations $A_n = A J_n$ of A (cf. [29]). Consider the two approximating equations*

$$u_n(t) = \phi(t) + \sum_{j=1}^{\infty} \int_0^t e^{(t-s)A} B^j J_n u_n(s) \, dw^j(s), \qquad (3.11)$$

$$v_n(t) = \phi(t) + \sum_{j=1}^{\infty} \int_0^t e^{(t-s)A} J_n B^j v_n(s) \, dw^j(s). \qquad (3.12)$$

They have unique solutions $u_n, v_n \in C_{\mathcal{F}}(\mathcal{X}) \cap L_{\mathcal{F}}^2(\mathcal{Y})$, both converging to the solution u of equation (3.1) in $C_{\mathcal{F}}(\mathcal{X}) \cap L_{\mathcal{F}}^2(\mathcal{Y})$.

Proof Step 1— Since $\|J_n\|_{L(\mathcal{X})} \leq 1$ (cf. [29]), from the property (3.5) it follows

$$\sum_{j=1}^{\infty} \int_0^\tau |B^j J_n e^{tA} x|_{\mathcal{X}}^2 \, dt = \sum_{j=1}^{\infty} \int_0^\tau |B^j e^{tA} J_n x|_{\mathcal{X}}^2 \, dt$$

$$\leq (\eta + c_2 \tau)|J_n x|_{\mathcal{X}}^2 \leq (\eta + c_2 \tau)|x|_{\mathcal{X}}^2,$$

and

$$\sum_{j=1}^{\infty} \int_0^\tau |J_n B^j e^{tA} x|_{\mathcal{X}}^2 \, dt \leq \sum_{j=1}^{\infty} \int_0^\tau |B^j e^{tA} x|_{\mathcal{X}}^2 \, dt \leq (\eta + c_2 \tau)|x|_{\mathcal{X}}^2.$$

Thus the new equations (3.11) and (3.12) satisfy the assumptions of Theorem 3.2, with the same constants. Hence, the existence and uniqueness follows from the previous results, and the bound (3.10) holds for equations (3.11) and (3.12) with a constant independent of n (this fact will be used below).

Step 2 — To prove the convergence, let $z_n = u - u_n$ and $y_n = u - v_n$. Then

$$z_n(t) = \sum_{j=1}^{\infty} \int_0^t e^{(t-s)A} B^j J_n z_n(s) \, dw^j(s) + \xi_n(t), \qquad (3.13)$$

$$y_n(t) = \sum_{j=1}^{\infty} \int_0^t e^{(t-s)A} J_n B^j y_n(s) \, dw^j(s) + \eta_n(t), \qquad (3.14)$$

where

$$\xi_n(t) = \sum_{j=1}^{\infty} \int_0^t e^{(t-s)A} B^j [u(s) - J_n u(s)] \, dw^j(s), \qquad (3.15)$$

$$\eta_n(t) = (I - J_n) \sum_{j=1}^{\infty} \int_0^t e^{(t-s)A} B^j u(s) \, dw^j(s). \qquad (3.16)$$

Since the semigroup e^{tA} restricted to $\mathcal{Y}$ is still a contraction semigroup in the graph topology (strightforward proof), $J_n x \to x$ in $\mathcal{Y}$ for all $x \in \mathcal{Y}$. Hence, by Lebesgue Dominated Convergence Theorem, $J_n u \to u$ in $L^2_{\mathcal{F}}(\mathcal{Y})$. By assumption (3.9) we then have $||J_n u - u||_{L^2_{\mathcal{F}}(\mathcal{Y}(B))} \to 0$, and this implies $\xi_n \to 0$ in $L^2_{\mathcal{F}}(\mathcal{Y})$ by the estimate of Lemma 3.1. From the result of [19] we also have $\xi_n \to 0$ in $C_{\mathcal{F}}(\mathcal{X})$. Hence, finally, $z_n \to 0$ in $L^2_{\mathcal{F}}(\mathcal{Y}) \cap C_{\mathcal{F}}(\mathcal{X})$ by the estimate (3.10).

Similarly, since

$$\sum_{j=1}^{\infty} \int_0^t e^{(t-s)A} B^j u(s) \, dw^j(s) \qquad (3.17)$$

is an element of $L^2_{\mathcal{F}}(\mathcal{Y}) \cap C_{\mathcal{F}}(\mathcal{X})$, and J_n strongly converges to the identity both in $\mathcal{X}$ and in $\mathcal{Y}$, by Lebesgue Dominated Convergence Theorem $\eta_n \to 0$ in $L^2_{\mathcal{F}}(\mathcal{Y}) \cap C_{\mathcal{F}}(\mathcal{X})$, and then $y_n \to 0$ as z_n above. This completes the proof of the Lemma.

3.4 An Extension

For applications to *non-homogeneous boundary value problems* we shall need a generalization of the previous results. The novelty is that a second pair of spaces, $\mathcal{X}_0$ and $\mathcal{Y}_0$, larger than $\mathcal{X}$ and $\mathcal{Y}$ respectively, are introduced. The idea is that $\mathcal{X}$ and $\mathcal{Y}$ are closed subspaces of $\mathcal{X}_0$ and $\mathcal{Y}_0$; in applications $\mathcal{X}_0$ and $\mathcal{Y}_0$ will be certain Sobolev spaces, while $\mathcal{X}$ and $\mathcal{Y}$ will be subspaces of them defined by some homogeneous boundary condition.

Theorem 3.5 *Let $\mathcal{Y} \subset \mathcal{X}$ and $A : D(A) \subset \mathcal{X} \to \mathcal{X}$ be as in Lemma 3.1.*
Let $\mathcal{Y}_0$ be another real separable Hilbert space, with $\mathcal{Y} \subset \mathcal{Y}_0$, and let B^j, $j \in N$, be linear operators satisfying the following assumptions:

$$B^j \in L(\mathcal{Y}_0, \mathcal{X}); \qquad (3.18)$$

$$\sum_{j=1}^{\infty} \int_0^{\tau} |B^j e^{tA} x|^2_{\mathcal{X}} \, dt \leq (\eta + c_2 \tau)|x|^2_{\mathcal{X}}, \tag{3.19}$$

for all $x \in D(A)$ and $\tau \in [0, T]$, and for some constants $\eta \in (0, 1)$ and $c_2 > 0$.

Let $\phi \in L^2_{\mathcal{F}}(\mathcal{Y}_{0,(B)})$, where the space $L^2_{\mathcal{F}}(\mathcal{Y}_{0,(B)})$ is defined similarly to $L^2_{\mathcal{F}}(\mathcal{Y}_{(B)})$. Then the equation

$$u(t) = \phi(t) + \sum_{j=1}^{\infty} \int_0^t e^{(t-s)A} B^j u(s) \, dw^j(s) \tag{3.20}$$

has a unique solution u in $L^2_{\mathcal{F}}(\mathcal{Y}_{0,(B)})$, and

$$||u||^2_{L^2_{\mathcal{F}}(\mathcal{Y}_{0,(B)})} \leq c ||\phi||^2_{L^2_{\mathcal{F}}(\mathcal{Y}_{0,(B)})}$$

for some constant $c > 0$ independent of ϕ.

The proof is the same of that of Theorem 3.2, without any change. Similarly to Corollary 3.3, here we have:

Corollary 3.6. Let $\mathcal{Y} \subset \mathcal{X}$ and $A : D(A) \subset \mathcal{X} \to \mathcal{X}$ be as in Corollary 3.3. Let $\mathcal{Y}_0 \subset \mathcal{X}_0$ be two other real separable Hilbert spaces, with $\mathcal{Y} \subset \mathcal{Y}_0$, and $\mathcal{X} \subset \mathcal{X}_0$. Let B^j, $j \in \mathbf{N}$, be linear operators satisfying the following assumptions:

$$B^j \in L(\mathcal{Y}_0, \mathcal{X}); \tag{3.21}$$

$$\frac{1}{2} \sum_{j=1}^{\infty} |B^j u|^2_{\mathcal{X}} \leq -\eta < Au, u >_{\mathcal{X}} + \lambda |u|^2_{\mathcal{X}}, \quad u \in D(A), \tag{3.22}$$

and

$$\sum_{j=1}^{\infty} |B^j u|^2_{\mathcal{X}} \leq c_3 |u|^2_{\mathcal{Y}_0}, \quad u \in \mathcal{Y}_0, \tag{3.23}$$

for some constant $\eta \in (0, 1)$, $\lambda > 0$, $c_3 > 0$.

Then, for all $\phi \in C_{\mathcal{F}}(\mathcal{X}_0) \cap L^2_{\mathcal{F}}(\mathcal{Y}_0)$ there exists a unique solution u of equation (3.20) in $C_{\mathcal{F}}(\mathcal{X}_0) \cap L^2_{\mathcal{F}}(\mathcal{Y}_0)$, and

$$|u|^2_{C_{\mathcal{F}}(\mathcal{X}_0)} + |u|^2_{L^2_{\mathcal{F}}(\mathcal{Y}_0)} \leq c(|\phi|^2_{C_{\mathcal{F}}(\mathcal{X}_0)} + |\phi|^2_{L^2_{\mathcal{F}}(\mathcal{Y}_0)})$$

for some constant $c > 0$ independent of ϕ.

3.5 Hilbert Scales and Interpolation of Random Field Spaces

We say that a family of real separable Hilbert spaces $(H) = \{H_\alpha; \alpha \in R\}$ is a *Hilbert scale* if

$$H_\beta \subset H_\alpha, \quad \alpha < \beta, \tag{3.24}$$

$$[H_\alpha, H_\beta]_\theta = H_{\theta\beta+(1-\theta)\alpha}, \quad \alpha < \beta, \theta \in (0, 1), \tag{3.25}$$

where $[X, Y]_\theta$ are the *complex interpolation spaces*, defined for instance in [25], [36]. In (3.25) we only assume that the norms in $[H_\alpha, H_\beta]_\theta$ and $H_{\theta\beta+(1-\theta)\alpha}$ are equivalent.

The way we use (3.25) is by the basic interpolation property (cf. [36], [25]): let $(H) = \{H_\alpha; \alpha \in \mathbf{R}\}$ and $(V) = \{V_\alpha; \alpha \in \mathbf{R}\}$ be two Hilbert scales, and

(IP) let T be a bounded linear operator from H_α to V_α for some $\alpha \in \mathbf{R}$. Assume that for some $\beta > \alpha$, $T(H_\beta) \subset V_\beta$, and $T|_{H_\beta}$ is a bounded linear operator from H_β to V_β. Then, for all $\theta \in (0, 1)$, $T(H_{\theta\beta+(1-\theta)\alpha}) \subset V_{\theta\beta+(1-\theta)\alpha}$, and $T|_{H_{\theta\beta+(1-\theta)\alpha}}$ is a bounded linear operator from $H_{\theta\beta+(1-\theta)\alpha}$ to $V_{\theta\beta+(1-\theta)\alpha}$.

As it is customary, we shall denote the restrictions of such operators T by T itself.

Properties (3.25) and (IP) carry over in various ways to the corresponding spaces $L^2(\mathcal{F}_t; H_\alpha)$, $C_{\mathcal{F}}(H_\alpha)$ and $L^2_{\mathcal{F}}(H_\alpha)$:

Theorem 3.7 *For $\alpha < \beta$ and $\theta \in (0, 1)$,*

i) $$[L^2(\mathcal{F}_t; H_\alpha), L^2(\mathcal{F}_t; H_\beta)]_\theta = L^2(\mathcal{F}_t; H_{\theta\beta+(1-\theta)\alpha}); \tag{3.26}$$

ii) $$[L^2_{\mathcal{F}}(H_\alpha), L^2_{\mathcal{F}}(H_\beta)]_\theta = L^2_{\mathcal{F}}(H_{\theta\beta+(1-\theta)\alpha}); \tag{3.27}$$

iii) if T is a bounded linear operator in the following spaces

$$T : L^2(\mathcal{F}_0; H_\alpha) \to L^2_{\mathcal{F}}(H_\alpha)$$

$$T : L^2(\mathcal{F}_0; H_\beta) \to L^2_{\mathcal{F}}(H_\beta),$$

then T is a bounded linear operator in

$$T : L^2(\mathcal{F}_0; H_{\theta\beta+(1-\theta)\alpha}) \to L^2_{\mathcal{F}}(H_{\theta\beta+(1-\theta)\alpha});$$

iv) similarly, if T is bounded linear in

$$T : L^2(\mathcal{F}_0; H_\alpha) \to C_{\mathcal{F}}(H_\alpha)$$

$$T : L^2(\mathcal{F}_0; H_\beta) \to C_{\mathcal{F}}(H_\beta),$$

then T is a bounded linear operator in

$$T : L^2(\mathcal{F}_0; H_{\theta\beta+(1-\theta)\alpha}) \to C_{\mathcal{F}}(H_{\theta\beta+(1-\theta)\alpha}).$$

Proof — i) and ii) are particular case of a classical result (cf. [36], Th. 1.18.4): if $(E, \mathcal{E}, \rho)$ is a σ-finite measure space, and (X, Y) is an interpolation couple, then

$$[L^2(E, \mathcal{E}, \rho; X), L^2(E, \mathcal{E}, \rho; Y)]_\theta = L^2(E, \mathcal{E}, \rho; [X, Y]_\theta)$$

(in ii) we take $E = \Omega \times [0, T]$, and $\mathcal{E}$ the σ-field of progressively measurable events (cf. [33], 1.5.11).

Part iii) follows from i) and ii), according to the interpolation property (IP). Finally, iv) would follow from the identity

$$[C_{\mathcal{F}}(H_\alpha), C_{\mathcal{F}}(H_\beta)]_\theta = C_{\mathcal{F}}(H_{\theta\beta+(1-\theta)\alpha}),$$

but its proof is involved because of the adaptation. Thus we limit ourself to prove explicity the interpolation property iv). The assumptions of iv) imply those of iii), so that, if $u_0 \in L^2(\mathcal{F}_0; H_{\theta\beta+(1-\theta)\alpha})$, then $Tu_0 \in L^2_{\mathcal{F}}(H_{\theta\beta+(1-\theta)\alpha})$. Hence Tu_0 is a progressively measurable process with values in $H_{\theta\beta+(1-\theta)\alpha}$. Moreover, it is well known ([25], Th. 1.14.2) that

$$[C([0, T]; H_\alpha), C([0, T]; H_\beta)]_\theta = C([0, T]; H_{\theta\beta+(1-\theta)\alpha}).$$

Hence, by the classical result mentioned above,

$$[L^2(\Omega, \mathcal{F}_T, P; C([0, T]; H_\alpha)), L^2(\Omega, \mathcal{F}_T, P; C([0, T]; H_\beta))]_\theta$$

$$= L^2(\Omega, \mathcal{F}_T, P; C([0, T]; H_{\theta\beta+(1-\theta)\alpha})).$$

Therefore

$$Tu_0 \in L^2(\Omega, \mathcal{F}_T, P; C([0, T]; H_{\theta\beta+(1-\theta)\alpha})),$$

and depend continuously on $u_0 \in L^2(\mathcal{F}_0; H_{\theta\beta+(1-\theta)\alpha})$. Together with the predictability, this yields the desired conclusion.

4   Regularity in Bounded Domains: Some Counterexamples

In this section we show that the differential operators of stochastic parabolic equations in bounded domains must be subject to certain *geometric restrictions* if one would develop a regularity theory similar to the deterministic one. The regularity theory of deterministic linear partial differential equations of parabolic type (and other types as well) tells us that, besides a considerable number of details related to the definition of the proper function spaces and the compatibility relations, the smoother are the data (initial data, boundary values, forcing terms) the smoother are the solution, with a rather elegant property of shift of the order of regularity. A central aim of this paper is to extend as much as possible this useful principle to stochastic parabolic equations. However, as we shall see in this section, in order to develop this programm we have to impose certain geometric restrictions on the differential operators (new with respect to the deterministic case). We shall clarify these ideas with some elementary counterexample in one space dimension, giving only a partial idea on the general multidimensional case. However, as partial as these counterexamples could be, they should convince the reader that the restrictions imposed in sections 5 and 6 at the level of sufficient conditions for a regularity theory are essentially necessary.

Consider first the one-dimensional bilinear equation with homogeneous Dirichlet boundary conditions in the bounded domain $D = (0, 1)$

$$
\begin{cases}
du(t, x) = \frac{\partial^2 u}{\partial x^2}(t, x)dt + b(x)\frac{\partial u}{\partial x}(t, x)\, dw(t), \\
u = 0, \quad t \in [0, T], x = 0, 1, \\
u(0, x) = u_0(x), \quad x \in [0, 1],
\end{cases}
\tag{4.1}
$$

where $x \in [0, 1]$, and $b(.)$ is a smooth function on $[0, 1]$. Assume that $b(x)^2 \leq 2\eta$ for all $x \in [0, 1]$ and for some $\eta \in (0, 1)$ (this is condition (3.8)). Assume $u_0 \in L^2(D)$. Here, by solution of (4.1) we mean a process $u \in C_{\mathcal{F}}(L^2(D)) \cap L^2_{\mathcal{F}}(H^1(D))$, with $u(0) = u_0$, which satisfies the boundary condition in (4.1) for a.e. (t, ω) ($H^1(D) \subset C(\overline{D})$, so that the boundary condition make sense), and the integral equation

$$
u(t, x) = u_0(x) + \int_0^t \frac{\partial^2 u}{\partial x^2}(s, x)ds + \int_0^t b(x)\frac{\partial u}{\partial x}(s, x)\, dw(s),
\tag{4.2}
$$

is satisfied, as an identity in $C_{\mathcal{F}}(H^{-1}(D))$, having defined the operations in a vector valued sense. One can show that this notion of solution is equivalent to the mild solution introduced above, with a suitable definition of A and B^j, but we omit the proof of this fact since it is not strictly necessary in the sequel.

Theorem 4.1 *If $b(0) \neq 0$ and $b(1) \neq 0$, then there exists no solutions (different from 0) of system (4.1) with the regularity $u \in C_{\mathcal{F}}(H^{1+s}(D)) \cap L^2_{\mathcal{F}}(H^{2+s}(D))$, $s > \frac{1}{2}$.*

Proof — Let u be a solution with the regularity stated above. Note that $u_0 \in H^{1+s}(D)$ a fortiori, so that the values of u_0 on the boundary are well defined. Under this regularity we can apply to each term of (4.2) the operator of trace on the boundary (point evaluation at 0 and 1, here). For instance, at $x = 0$ we have

$$
u(t, 0) = u_0(0) + \int_0^t \frac{\partial^2 u}{\partial x^2}(s, 0)ds + \int_0^t b(0)\frac{\partial u}{\partial x}(s, 0)\, dw(s).
$$

The first two terms vanish, while the third one, $\int_0^t \frac{\partial^2 u}{\partial x^2}(s, 0)ds$, is a process with absolutely continuous paths, P- a.s., then with quadratic variation equal to 0. Hence the quadratic variation of

$$\int_0^t b(0)\frac{\partial u}{\partial x}(s,0)\,dw(s)$$

is equal to 0. Thus, by the assumption on b, we obtain $\frac{\partial u}{\partial x}|_{x=0} = 0$, for a.e. (t,ω), and similarly for $\frac{\partial u}{\partial x}|_{x=1}$. This allows us to extend u to a solution of a stochastic parabolic equation over R; such solution has compact support (the closure of D), a property that solutions to parabolic equations in the full space cannot have, unless vanishing identically. To develop the details, let us set $\tilde{u} = u$ for $x \in [0,1]$, and $\tilde{u} = 0$ for $x \notin [0,1]$. We have $\tilde{u} \in C_{\mathcal{F}}(L^2(\mathbf{R})) \cap L^2_{\mathcal{F}}(H^1(\mathbf{R}))$ (at least), and one can easily check that $\tilde{u}$ is a solution of the equation

$$\begin{cases} d\tilde{u}(t,x) = \frac{\partial^2 \tilde{u}}{\partial x^2}(t,x)dt + \tilde{b}(x)\frac{\partial \tilde{u}}{\partial x}(t,x)\,dw(t), \\ \tilde{u}(0,x) = \tilde{u}_0(x), \quad x \in R, \end{cases} \tag{4.3}$$

where $x \in R$, $\tilde{u}_0$ is the extension of u_0 outside $[0,1]$ similar to $\tilde{u}$, and $\tilde{b}$ extends b outside $[0,1]$ in a smooth way, preserving the coercivity condition $\tilde{b}(x)^2 \le 2\tilde{\eta}$ for all $x \in R$ and for some $\tilde{\eta} \in (0,1)$. Equation (4.3) has a unique solution in $C_{\mathcal{F}}(L^2(\mathbf{R})) \cap L^2_{\mathcal{F}}(H^1(\mathbf{R}))$, which is also in $C_{\mathcal{F}}(H^n(\mathbf{R}))$ for all n (see [30], or section 5.2), by the regularity of the coefficients. From the stochastic characteristic representation formula (cf. [37]) we have

$$\tilde{u}(t,x) = v(t, \phi_t^{-1}(x)), \tag{4.4}$$

where v satisfies the parabolic equation with random coefficients

$$\begin{aligned} \frac{\partial v}{\partial t} = {}& [1 - \frac{1}{2}\tilde{b}(\phi_t(x))^2][\frac{\partial \phi_t^{-1}}{\partial x}(\phi_t(x))]^2 \frac{\partial^2 v}{\partial x^2} \\ & + [1 - \frac{1}{2}\tilde{b}(\phi_t(x))^2]\frac{\partial^2 \phi_t^{-1}}{\partial x^2}(\phi_t(x))\frac{\partial v}{\partial x} \\ & - \frac{1}{2}\tilde{b}(\phi_t(x))\frac{\partial \tilde{b}}{\partial x}(\phi_t(x))\frac{\partial \phi_t^{-1}}{\partial x}(\phi_t(x))\frac{\partial v}{\partial x}, \end{aligned}$$

with

$$v(0,x) = \tilde{u}_0(x),$$

and $\phi_t(x)$ is the stochastic flow associated to the equation

$$\begin{cases} d\xi_t = \frac{1}{2}\tilde{b}(\xi_t)\frac{\partial \tilde{b}}{\partial x}(\xi_t)\,dt - \tilde{b}(\xi_t)\,dw(t), \quad t \ge 0, \\ \xi_0 = x. \end{cases} \tag{4.5}$$

To clarify the argument, assume $\tilde{b}(x) = 1$ for all $x \in \mathbf{R}$ (the general case is similar). Then v satisfies the equation

$$\begin{cases} \frac{\partial v}{\partial t} = \frac{1}{2}\frac{\partial^2 v}{\partial x^2}, \\ v(0, x) = \tilde{u}_0(x), \end{cases} \qquad (4.6)$$

$x \in R$, and the solution of (4.5) is $\xi_t = x - w(t)$, so that

$$\phi_t^{-1}(x) = x + w(t).$$

Now, let $x_0 \notin [0, 1]$ and $t_0 > 0$ arbitrary. By construction $\tilde{u}(t_0, x_0) = 0$ P-a.s., hence $v(t_0, x_0 + w(t_0)) = 0$ P-a.s. Since the support of $x_0 + w(t_0)$ is the whole real line, the last fact implies $v(t_0, .) = 0$, whence finally $\tilde{u}(t_0, .) = 0$, completing the proof by the arbitrariety of t_0.

Remark — Let us discuss the intuitive reason for this lack of smoothness. The irregularity of u appears only near the boundary of D: in the interior it reflects the regularity of the data. At the boundary, the flux of u (given by $b(0)\frac{\partial u}{\partial x}(t, 0)\frac{dw}{dt}(t)$ at $x = 0$, for instance) has the speed of white noise. The solution is forced to be equal to 0 at $x = 0$; thus, when the (very fast) flux points outward, there is a tendence to create a discontinuity between the value $u = 0$ at $x = 0$ and the values of u close to $x = 0$; when the flux points inward, the value $u = 0$ at $x = 0$ is approximately extended to a small region inside $(0, 1)$, close to $x = 0$. This process is very fast, and balance exactly the speed of the diffusion due to the term $\frac{\partial^2 u}{\partial x^2}$, which otherwise would have a smoothing effect. Thus the solution oscillates very fast near the boundary.

Higher regularity of solutions imposes further constraints (each step of two degrees of regularity yields a new constraint). For instance we have:

Theorem 4.2 *Assume* $b = 0$ *for* $x = 0, 1$. *If* $\frac{\partial^2 b}{\partial x^2}(0) \neq 0$ *and* $\frac{\partial^2 b}{\partial x^2}(1) \neq 0$, *then there exists no solutions (different from 0) of system (4.1) with the regularity* $u \in C_{\mathcal{F}}(H^{3+s}(D)) \cap L^2_{\mathcal{F}}(H^{4+s}(D))$, $s > \frac{1}{2}$.

Proof — Taking the trace at $x = 0, 1$ in equation (4.2) and using the new assumption that $b = 0$ for $x = 0, 1$, we get

$$\frac{\partial^2 u}{\partial x^2} = 0 \qquad (4.7)$$

for $x = 0, 1$ and for a.e. (t, ω). By (4.2) again, we have

$$\frac{\partial^2 u}{\partial x^2}(t, x) = \frac{\partial^2 u_0}{\partial x^2}(x) + \frac{\partial^2}{\partial x^2}\int_0^t \frac{\partial^2 u}{\partial x^2}ds + \int_0^t \frac{\partial^2}{\partial x^2}b\frac{\partial u}{\partial x}dw(s), \qquad (4.8)$$

and computing this identity for $x = 0, 1$ we obtain (using again the quadratic variation argument of the previous proof)

$$\frac{\partial^2 b}{\partial x^2}\frac{\partial u}{\partial x} + 2\frac{\partial b}{\partial x}\frac{\partial^2 u}{\partial x^2} + b\frac{\partial^3 u}{\partial x^3} = 0 \qquad (4.9)$$

for $x = 0, 1$ and for a.e. (t, ω). The last two terms vanish, by (4.7) and the assumption on b, so that, by the assumption on $\frac{\partial^2 b}{\partial x^2}$, we have

$$\frac{\partial u}{\partial x} = 0$$

for $x = 0, 1$ and for a.e. (t, ω). Thus, the last part of the proof of Theorem 4.1 can be repeated, completing the proof of the present Theorem.

We can summarize the (partial) informations given by the two previous Theorems as follows. Consider a more general stochastic parabolic equation in a bounded domain $D \subset \mathbf{R}^d$, with homogeneous Dirichlet boundary condition on the boundary Γ of D, of the form

$$\begin{cases} du(t, x) = (\mathcal{A}u)(t, x)dt + \sum (B^j u)(t, x) \, dw^j(t), \\ u = 0, \quad t \in [0, T], x \in \Gamma, \\ u(0, x) = u_0(x), \quad x \in D. \end{cases} \quad (4.10)$$

Here $\mathcal{A}$ and B^j are differential operators of order two and one respectively, the form of which will be specified in section 5.2. Consider the following condition on B^j:

$$B^j u|_\Gamma = 0, \quad \text{for all} \quad u \in H^{1+s}(D) \quad \text{such that} \quad u|_\Gamma = 0, \quad (4.11)$$

where $s > \frac{1}{2}$. If B^j has the form

$$B^j u = b(x) \cdot \nabla u \quad (4.12)$$

for a smooth vector field b, condition (4.11) is equivalent to:

$$b(x) \quad \text{tangent to} \quad \Gamma \quad \text{for a.e.} \quad x \in \Gamma \quad (4.13)$$

(indeed, the condition $u|_\Gamma = 0$ in (4.11) implies that ∇u is normal to Γ). We have the following extension of Theorem 4.1. Since this result is not needed in the sequel, and its role is only indicative, we do not give the lenghtly complete proof.

Theorem 4.3 *If condition (4.11), or (4.13), is not satisfied, there exists no solutions (different from 0) of system (4.10) with the regularity $u \in C_{\mathcal{F}}(H^{1+s}(D)) \cap L^2_{\mathcal{F}}(H^{2+s}(D))$, $s > \frac{1}{2}$, and (at least) with nonnegative u_0.*

Outline of the Proof — If condition (4.11) is not satisfied, there is a ball S centered at a point of the boundary such that $S \cap \Gamma$ is diffeomorphic to a ball of $\mathcal{R}^{d-1}$, and $b(x)$ is not tangent to Γ, for a.e. $x \in S \cap \Gamma$. Arguing as in the proof of Theorem 4.1, we can extend the domain D to $D \cup S$, and extend the solution u of (3.17) to a solution $\tilde{u}$ of the extended problem on $D \cup S$. However, $\tilde{u}$ has support in D, thus strictly contained in $D \cup S$. We obtain that u is 0 everywhere, by a more complex argument with respect to that used in Theorem 4.1, based on a Feynmann-Kac representation of the form

$$\tilde{u}(t, x) = \hat{E}\tilde{u}_0(\xi_{t \wedge \tau(x)}(x)),$$

where to $\tilde{u}_0$ is given the value 0 on the boundary of $D \cup S$, $\xi_t(x)$ is a diffusion starting from x, driven by a pair of Wiener processes, one of them being the Wiener process of equation (4.10) and the other being an auxiliary Wiener process with respect to which the mean $\hat{E}$ is taken, and $\tau(x)$ is the first exit time of $\xi_t(x)$ from $D \cup S$ (cf. [16] for the details). Since the diffusion $\xi_t(x)$ is non-degenerate, we can apply the same support argument as in the proof of Theorem 4.1 to the process $\tilde{u}_0(\xi_{t \wedge \tau(x)}(x))$. Under the additional assumption that u_0 is non-negative, the mean $\hat{E}$ does not affect the argument, and the proof can be completed.

One can see that the obstruction discussed in Theorem 4.2 is related to the abstract condition:

$$\mathcal{A} B^j u|_\Gamma = 0 \quad \text{for all} \quad u \in H^{3+s}(D), u|_\Gamma = 0 \quad \text{such that} \quad \mathcal{A} u|_\Gamma = 0, \qquad (4.14)$$

where $s > \frac{1}{2}$. One can also formulate the conditions for higher order of regularity, but we omit the details here.

To conclude this section, as a further counterexample, we discuss an elementary Neumann boundary value problem:

$$\begin{cases} du(t, x) = \frac{\partial^2 u}{\partial x^2}(t, x)dt + b\frac{\partial u}{\partial x}(t, x)\, dw(t), \\ \frac{\partial u}{\partial x} = 0, \quad t \in [0, T], x = 0, 1, \\ u(0, x) = u_0(x), \quad x \in [0, 1], \end{cases} \qquad (4.15)$$

where $x \in [0, 1]$, and $b \neq 0$ is a constant (a smooth function $b(x)$ on $[0, 1]$ can be dealt with, by some modifications).

Theorem 4.4 *If $b \neq 0$, then there exists no solutions (different from 0) of system (4.15) with the regularity $u \in C_{\mathcal{F}}(H^{1+s}(D)) \cap L^2_{\mathcal{F}}(H^{2+s}(D))$, $s > \frac{3}{2}$.*

Proof — Taking the x-derivative at $x = 0, 1$ in equation (4.2), by the argument of the proof of Theorem 4.1 we obtain

$$\frac{\partial}{\partial x}(b\frac{\partial u}{\partial x}(t, x)) = 0 \qquad (4.16)$$

for $x = 0, 1$ and for a.e. (t, ω). Hence

$$\frac{\partial^2 u}{\partial x^2} = 0 \qquad (4.17)$$

for $x = 0, 1$ and for a.e. (t, ω). Then the function $v = \frac{\partial u}{\partial x}$ satisfies the system

$$\begin{cases} dv(t, x) = \frac{\partial^2 v}{\partial x^2}(t, x)dt + b\frac{\partial v}{\partial x}(t, x)\, dw(t), \\ v = 0, \frac{\partial v}{\partial x} = 0, \quad t \in [0, T], x = 0, 1, \\ v(0, x) = \frac{\partial u_0}{\partial x}(x), \quad x \in [0, 1]. \end{cases} \qquad (4.18)$$

The final argument of the proof of Theorem 4.1 can be repeated, completing the proof of the Theorem.

5 Regularity Theory for Homogeneous Problems

5.1 Abstract Regularity Theory

The results of this section are based on Corollary 3.3. Let H be a real separable Hilbert space. Let $A : D(A) \subset H \to H$ be the infinitesimal generator of an *analytic semigroup* of negative type. Let

$$V_\alpha = D((-A)^{\alpha/2}), \quad \alpha \in \mathbf{R},$$

endowed with some norm $|x|_\alpha$ and inner product $< ., . >_\alpha$ equivalent to the graph topology (for $\alpha = 0$, which is excluded by the usual definitions of fractional powers, we simply set $V_0 = H$). Here $(-A)^\alpha$ denote the fractional powers of $(-A)$, defined for instance in [29], [34], [36]. The family $(V) = \{V_\alpha; \alpha \in \mathbf{R}\}$ form a Hilbert scale (cf. [36], Ch. 1). We note that

$$A : V_{\alpha+2} \to V_\alpha \quad \forall \alpha \in \mathbf{R}, \tag{5.1}$$

are linear bounded operators (with slight abuse of notation: for $\alpha > 0$ we mean the restriction of A to $V_{\alpha+2}$; for $\alpha < 0$ there exists a unique extension of A to a linear bounded operator from $V_{\alpha+2}$ to V_α). Moreover, for each $\alpha \in \mathbf{R}$, the operator given by (5.1) generates an analytic semigroup in V_α, always denoted by e^{tA}, $t \geq 0$. Thus, in each V_α, A fulfills the first conditions of Corollary 3.3.

Over the time interval $[0, T]$ we consider the following abstract equation

$$\begin{cases} du(t) = Au(t)\, dt + \sum_{j=1}^\infty B^j u(t)\, dw^j(t) \\ u(0) = u_0, \end{cases} \tag{5.2}$$

interpreted in the sequel always in the *mild form*

$$u(t) = e^{tA} u_0 + \sum_{j=1}^\infty \int_0^t e^{(t-s)A} B^j u(s)\, dw^j(s). \tag{5.3}$$

The linear unbounded operators B_j will be specified below.

Definition 5.1: *Given $\alpha \in \mathbf{R}$, assume that B^j are bounded linear operators from $V_{\alpha+1}$ to V_α. We say that problem (5.2) is well posed in V_α if for every $u_0 \in L^2(\mathcal{F}_0; V_\alpha)$ there exists a unique mild solution u of equation (5.3) in $C_{\mathcal{F}}(V_\alpha) \cap L^2_{\mathcal{F}}(V_{\alpha+1})$, and u depends continuously on u_0 in these topologies.*

Applying Corollary 3.3. in the space $\mathcal{X} = V_\alpha$ (so that $\mathcal{Y} = V_{\alpha+1}$) we readily have:

Theorem 5.2 *Let $\alpha \in \mathbf{R}$ be fixed. Let*

$$B^j : V_{\alpha+1} \to V_\alpha \qquad (5.4)$$

be bounded linear operators. Assume that for some Hilbert norms $|.|_\alpha$ and $|.|_{\alpha+1}$ equivalent to the graph topologies of V_α and $V_{\alpha+1}$ we have

$$\frac{1}{2} \sum_{j=1}^\infty |B^j u|_\alpha^2 \le -\eta < Au, u >_\alpha + \lambda |u|_\alpha^2, \quad u \in V_{\alpha+2}, \qquad (5.5)$$

and

$$\sum_{j=1}^\infty |B^j u|_\alpha^2 \le c |u|_{\alpha+1}^2, \quad u \in V_{\alpha+1}, \qquad (5.6)$$

for some constants $\eta \in (0, 1)$, $\lambda \ge 0$, and $c > 0$.
Then equation (5.2) is well posed in V_α.

Definition 5.3: *Given $\alpha \in \mathbf{R}$, assume that B^j are bounded linear operators from $V_{\alpha+1}$ to V_α. We say that problem (5.2) is well posed and coercive in V_α if the assumptions of Theorem 5.2 are satisfied (so that problem (5.2) is well posed a fortiori).*

From the interpolation Theorem 3.7, we also have a result of well posedness in intermediate spaces, and the smoothing property typical of parabolic equations.

Theorem 5.4 *Let $\alpha < \beta$ be given real numbers. If (5.2) is well posed in V_α and V_β, then it is well posed in V_γ for all $\gamma \in [\alpha, \beta]$. Moreover, for every $u_0 \in L^2(\mathcal{F}_0; V_\alpha)$ and $\varepsilon \in (0, T)$, the solution u belongs to $C_{\mathcal{F}}(\varepsilon, T; V_\beta) \cap L_{\mathcal{F}}^2(\varepsilon, T; V_{\beta+1})$, and the mappings $u_0 \mapsto u|_{[\varepsilon, T]}$ are continuous from $L^2(\mathcal{F}_0; V_\alpha)$ to $C_{\mathcal{F}}(\varepsilon, T; V_\beta) \cap L_{\mathcal{F}}^2(\varepsilon, T; V_{\beta+1})$ for all $\varepsilon \in (0, T)$.*

Proof — The first part follows by the interpolation Theorem 3.7. To prove the second part, note that $u(t) \in L^2(\mathcal{F}_t; V_{\alpha+1})$ for a.e. $t \in [0, T]$ (because $u \in L_{\mathcal{F}}^2(0, T; V_{\alpha+1})$), so that we can take an arbitrarily small $t_0 > 0$ such that $u(t_0) \in L^2(\mathcal{F}_{t_0}; V_{\alpha+1})$. Therefore, we have $u|_{[t_0, T]} \in C_{\mathcal{F}}(t_0, T; V_{\alpha+1}) \cap L_{\mathcal{F}}^2(t_0, T; V_{\alpha+2})$. Repeating this argument a finite number of times we prove $u|_{[\varepsilon, T]} \in C_{\mathcal{F}}(\varepsilon, T; V_\beta) \cap L_{\mathcal{F}}^2(\varepsilon, T; V_{\beta+1})$ for all $\varepsilon \in (0, T)$. Finally, the mapping $u_0 \mapsto u|_{[\varepsilon, T]}$ is closed, as one can readily see, and everywhere defined, whence continuous.

To prove regularity properties of solutions of ((5.2) we simply shall try to apply these results. To this end we have to impose certain restrictions on the coefficients of the differential operators B^j, to fulfill the assumption (5.4), as discussed also by the counterexamples of section 4 (the restriction in (5.4) is due to the boundary conditions of V_α). Besides this, the main effort is concerned with the proof of inequality (5.5) (as already remarked in section 3.2, assumption (5.6) is always easily satisfied in our applications). For $\alpha = 0$, the inequality

$$\frac{1}{2} \sum_{j=1}^{\infty} |B^j u|_0^2 \leq -\eta < Au, u >_0 + \lambda |u|_0^2, \quad u \in V_2, \tag{5.7}$$

corresponds (of course if H is an L^2-space) to a joint ellipticity condition on the coefficients of the differential operators A and B^j, which is classical in the theory of parabolic stochastic equations. The natural conjecture is that the same ellipticity condition implies also (5.5) for all α. Although we do not have a complete proof that this is true, at least for second order parabolic problems we can show that (5.5) is fulfilled for all positive and negative integers α (namely, in all Sobolev spaces with integer exponent). The strategy to obtain this result is to prove (5.5) directly at the level of concrete differential operators in the cases $\alpha = 0$ and $\alpha = 1$ (i.e. in $L^2(D)$ and $H^1(D)$). Then, by means of the following abstract criteria, we can translate the results up and down of two orders of regularity, thus covering all integers α.

The following abstract results are improvements of conditions given in [16] and [4]. Roughly speaking, they say that, if the abstract problem (5.2) is well posed in a certain V_α, and if mild commutativity conditions on A and B^j are satisfied (always fulfilled in our applications), then (5.2) is well posed also in $V_{\alpha+2}$ and $V_{\alpha-2}$.

Lemma 5.5 *Assume:*

i) $B^j \in L(V_{\alpha+1}, V_\alpha)$, *and*

$$\frac{1}{2} \sum_{j=1}^{\infty} |B^j u|_\alpha^2 \leq -\eta < \overset{\circ}{A}u, u >_\alpha + \lambda |u|_\alpha^2, \quad u \in V_{\alpha+2}, \tag{5.8}$$

$$\sum_{j=1}^{\infty} |B^j u|_\alpha^2 \leq c_1 |u|_{\alpha+1}^2, \quad u \in V_{\alpha+1}, \tag{5.9}$$

for some constants $\eta \in (0, 1)$, $\lambda \geq 0$, and $c_1 > 0$, and for some norms $|.|_\alpha$ and $|.|_{\alpha+1}$ equivalent to the graph topologies of V_α and $V_{\alpha+1}$;

ii) $B^j \in L(V_{\alpha+3}, V_{\alpha+2})$;

iii) Setting $L^j = AB^j - B^j A$, defined and bounded (a priori) from $V_{\alpha+3}$ to V_α, we have

$$\sum_{j=1}^{\infty} |L^j u|_\alpha^2 \leq c_2 |u|_{\alpha+2}^2, \quad u \in V_{\alpha+3}, \tag{5.10}$$

for some constant $c_2 > 0$, and for some norm $|.|_{\alpha+2}$ equivalent to the graph topology of $V_{\alpha+2}$.

Endow $V_{\alpha+2}$ with the graph norm with respect to V_α: $\|u\|_{\alpha+2} := |Au|_\alpha$. This norm is equivalent to $|u|_{\alpha+2}$. Denote by $<< .,. >>_{\alpha+2}$ the inner product corresponding to $\|u\|_{\alpha+2}$. Similarly, endow $V_{\alpha+3}$ with the graph norm $\|u\|_{\alpha+3} := |Au|_{\alpha+1}$ with respect to $V_{\alpha+1}$.

Then we have

$$\frac{1}{2}\sum_{j=1}^{\infty}||B^j u||^2_{\alpha+2} \leq -\tilde{\eta} << Au, u >>_{\alpha+2} +\tilde{\lambda}||u||^2_{\alpha+2}, \quad u \in V_{\alpha+4}, \tag{5.11}$$

$$\sum_{j=1}^{\infty}||B^j u||^2_{\alpha+2} \leq c||u||^2_{\alpha+3}, \quad u \in V_{\alpha+3}, \tag{5.12}$$

for some constants $\tilde{\eta} \in (0, 1)$, $\tilde{\lambda} \geq 0$, *and* $c > 0$. *Hence equation (5.2) is well posed in* $V_{\alpha+2}$.

Theorem 5.6 *If equation (5.2) is well posed and coercive in* V_α (*cf. Definition 5.3*), *and conditions ii)-iii) of the previous Lemma hold true, then it is well posed and coercive in* $V_{\alpha+2}$.

Proof of Lemma 5.5 — For all $u \in V_{\alpha+4}$ we have $B^j u \in V_{\alpha+2}$ and

$$\frac{1}{2}\sum_{j=1}^{\infty}||B^j u||^2_{\alpha+2} = \frac{1}{2}\sum_{j=1}^{\infty}|AB^j u|^2_\alpha = \frac{1}{2}\sum_{j=1}^{\infty}|B^j Au + L^j u|^2_\alpha$$

$$\leq \frac{1+\varepsilon}{2}\sum_{j=1}^{\infty}|B^j Au|^2_\alpha + c(\varepsilon)\sum_{j=1}^{\infty}|L^j u|^2_\alpha$$

$$\leq -\frac{1+\varepsilon}{2}\eta < AAu, Au >_\alpha +\frac{1+\varepsilon}{2}\lambda|Au|^2_\alpha + c(\varepsilon)c_2|u|^2_{\alpha+2}.$$

Inequality (5.11) readily follows from this estimate. The proof of (5.12) is completely analogous, and is omitted. The proof is complete.

Remark — In concrete applications it is rather cumbersome to check assumption (5.10) in the case of infinitely many nonzero operators L^j. For this reason a simple regularity theory can be developed only in the case when only a finite number of B^j are different from zero. Nevertheless we stated all the abstract results in the general case in order to make the role of the various assumptions as transparent as possible.

Lemma 5.7 *Assume:*

i) $B^j \in L(V_{\alpha+1}, V_\alpha)$, *and*

$$\frac{1}{2}\sum_{j=1}^{\infty}|B^j u|^2_\alpha \leq -\eta < Au, u >_\alpha +\lambda|u|^2_\alpha, \quad u \in V_{\alpha+2}, \tag{5.13}$$

$$\sum_{j=1}^{\infty}|B^j u|^2_\alpha \leq c_1|u|^2_{\alpha+1}, \quad u \in V_{\alpha+1}, \tag{5.14}$$

for some constants $\eta \in (0, 1)$, $\lambda \geq 0$, *and* $c_1 > 0$, *and for some norms* $|.|_\alpha$ *and* $|.|_{\alpha+1}$
equivalent to the graph topologies of V_α *and* $V_{\alpha+1}$;

ii) $B^j \in L(V_{\alpha-1}, V_{\alpha-2})$;

iii) *Setting* $L^j = AB^j - B^j A$, *defined and bounded (a priori) from* $V_{\alpha+1}$ *to* $V_{\alpha-2}$, *we have*

$$\sum_{j=1}^{\infty} |L^j u|_{\alpha-2}^2 \leq c_2 |u|_\alpha^2, \quad u \in V_{\alpha+1}, \tag{5.15}$$

for some constant $c_2 > 0$, *and for some norm* $|.|_{\alpha-2}$ *equivalent to the graph topology of* $V_{\alpha-2}$.

Endow $V_{\alpha-2}$ with the graph norm with respect to V_α: $\|u\|_{\alpha-2} := |A^{-1}u|_\alpha$. This norm is equivalent to $|u|_{\alpha-2}$. Denote by $\ll .,. \gg_{\alpha-2}$ the inner product corresponding to $\|u\|_{\alpha-2}$. Similarly, endow $V_{\alpha-1}$ with the graph norm $\|u\|_{\alpha-1} = |A^{-1}u|_{\alpha+1}$ with respect to $V_{\alpha+1}$.
Then we have

$$\frac{1}{2} \sum_{j=1}^{\infty} \|B^j u\|_{\alpha-2}^2 \leq -\tilde{\eta} \ll Au, u \gg_{\alpha-2} + \tilde{\lambda}\|u\|_{\alpha-2}^2, \quad u \in V_\alpha, \tag{5.16}$$

$$\sum_{j=1}^{\infty} \|B^j u\|_{\alpha-2}^2 \leq c\|u\|_{\alpha-1}^2, \quad u \in V_{\alpha-1}, \tag{5.17}$$

for some constants $\tilde{\eta} \in (0, 1)$, $\tilde{\lambda} \geq 0$, *and* $c > 0$. *Hence equation (5.2) is well posed in* $V_{\alpha-2}$.

Theorem 5.8 *If equation (5.2) is well posed and coercive in* V_α *(cf. Definition 5.3), and conditions ii)-iii) of the previous Lemma hold true, then it is well posed and coercive in* $V_{\alpha-2}$.

Proof of Lemma 5.7. Step 1 — For all $u \in V_\alpha$ we have

$$A^{-1}B^j u = B^j A^{-1}u - A^{-1}L^j A^{-1}u. \tag{5.18}$$

Indeed, $y = A^{-1}u \in V_{\alpha+2}$, so that (by iii)) $L^j y = AB^j y - B^j Ay$, whence $A^{-1}L^j y = B^j y - A^{-1}B^j Ay$, and the latter implies (5.18).

Step 2 — For all $u \in V_\alpha$ we have $B^j u \in V_{\alpha-2}$ and

$$\frac{1}{2} \sum_{j=1}^{\infty} \|B^j u\|_{\alpha-2}^2 = \frac{1}{2} \sum_{j=1}^{\infty} |A^{-1}B^j u|_\alpha^2 = \frac{1}{2} \sum_{j=1}^{\infty} |B^j A^{-1}u + A^{-1}L^j A^{-1}u|_\alpha^2$$

$$\leq \frac{1+\varepsilon}{2} \sum_{j=1}^{\infty} |B^j A^{-1}u|_\alpha^2 + c(\varepsilon) \sum_{j=1}^{\infty} |A^{-1}L^j A^{-1}u|_\alpha^2$$

$$\leq -\frac{1+\varepsilon}{2}\eta < AA^{-1}u, A^{-1}u >_\alpha +\frac{1+\varepsilon}{2}\lambda|A^{-1}u|_\alpha^2 + c(\varepsilon)\sum_{j=1}^\infty |L^j A^{-1}u|_{\alpha-2}^2.$$

Inequality (5.16) readily follows from this estimate. The proof of (5.17) is completely analogous, and is omitted. The proof is complete.

5.2 Application to Equations in $\mathbf{R}^d$

Consider the following stochastic parabolic equation in $[0, T] \times \mathbf{R}^d$:

$$\begin{cases} du(t, x) = \mathcal{A}u(t, x)\, dt + \sum_{j=1}^n B^j u(t, x)\, dw^j(t), \\ u(0, x) = u_0(x), \quad x \in \mathbf{R}^d, \end{cases} \tag{5.19}$$

where $\mathcal{A}$ and B^j are the differential operators

$$\mathcal{A}u(x) = \sum_{i,j=1}^d \frac{\partial}{\partial x_j}\left(a_{ij}(x)\frac{\partial u(x)}{\partial x_i}\right) + \sum_{i=1}^d a_i(x)\frac{\partial u(x)}{\partial x_i} + a(x)u(x) \tag{5.20}$$

$$B^j u(x) = \sum_{i=1}^d b_i^j(x)\frac{\partial u(x)}{\partial x_i} + c^j(x)u(x) \tag{5.21}$$

with bounded C^∞ coefficients $a_{ij}(x), a_i(x), a(x)$ and $b_i^j(x), c^j(x)$, satisfying the usual joint ellipticity condition

$$\sum_{i,j=1}^d \left(a_{ij}(x) - \frac{1}{2}\sum_{k=1}^n b_i^k(x)b_j^k(x)\right)\xi_i\xi_j \geq \rho\sum_{i=1}^d \xi_i^2, \tag{5.22}$$

$(\xi_1, \ldots, \xi_d) \in \mathbf{R}^d, x \in \mathbf{R}^d$, for some constant $\rho > 0$. In particular, this implies that the operator $\mathcal{A}$ is uniformly strongly elliptic.

Since we deal with stochastic equations over finite time horizon, we can always shift the operator $\mathcal{A}$ by a multiple λ of the identity (by considering the new variable $v(t) = e^{\lambda t}u(t)$), which amounts to change the coefficient $a(x)$ in $a(x) + \lambda$. In this way, by a suitable choice of λ, we can always assume that the semigroup generated by $\mathcal{A}$ (in the spaces specified in the sequel) is of negative type. We shall not repeat this comment in each application.

Concerning the fact that we take only a finite number of operators B^j in equation (5.19), see the remark at the end of the subsection.

Condition (5.22) implies that there exists $\eta \in (0, 1)$ such that

$$\frac{1}{2}\sum_{k=1}^n \sum_{i,j=1}^d b_i^k(x)b_j^k(x)\xi_i\xi_j \leq \eta\sum_{i,j=1}^d a_{ij}(x)\xi_i\xi_j, \tag{5.23}$$

$(\xi_1, \ldots, \xi_d) \in \mathbf{R}^d, x \in \mathbf{R}^d$. Let $A : H^2(\mathbf{R}^d) \to L^2(\mathbf{R}^d)$ be defined by $Au = \mathcal{A}u$. Then, by (5.23) and an elementary application of integration by parts, we see that the coercivity condition (5.7) is satisfied in $L^2(\mathbf{R}^d)$. By (5.7) and Theorem 5.2 it follows that equation (5.19) is well posed in $L^2(\mathbf{R}^d)$. It is trivial to check that all the conditions of Theorems 5.6 and 5.8 are satisfied, for all $n = 2k$, $k \in \mathbf{Z}$. In particular, note that the operators L^k are second order differential operators, with higher order part given by

$$\sum_{i,j,h} \left[(a_{i,j} + a_{j,i}) \frac{\partial b_h^k}{\partial x_i} \frac{\partial^2}{\partial x_i \partial x_h} - b_h^k \frac{\partial a_{i,j}}{\partial x_h} \frac{\partial^2}{\partial x_i \partial x_j} \right]. \tag{5.24}$$

Hence, all the operators L^k are bounded from $V_{\alpha+2}$ in V_α, for each $\alpha \in \mathbf{R}$, since $V_\alpha = H^\alpha(\mathbf{R}^d)$ (assumptions (5.10) and (5.15) of Lemmas 5.5 and 5.7 are satisfied).

Thus (5.19) is well posed also in $H^n(\mathbf{R}^d)$, for all $n = 2k$, $k \in \mathbf{Z}$. Finally, applying Theorem 5.4, we obtain that equation (5.19) is well posed in all Sobolev spaces $H^\alpha(\mathbf{R}^d)$, and that the regularizing property stated there holds true. Summarizing, we have:

Theorem 5.9 *For all $\alpha \in \mathbf{R}$, equation (5.19) is well posed in $H^\alpha(\mathbf{R}^d)$. Moreover, given $u_0 \in L^2(\mathcal{F}_0, H^\alpha(\mathbf{R}^d))$, the corresponding solution u satisfies*

$$u \in C_\mathcal{F}(\varepsilon, T; C^m(\mathbf{R}^d)), \quad \forall \varepsilon \in (0, T), \forall m \in \mathbf{N},$$

and depends continuously on u_0 between $L^2(\mathcal{F}_0, H^\alpha(\mathbf{R}^d))$ and $C_\mathcal{F}(\varepsilon, T; C^m(\mathbf{R}^d))$.

Of course we substituted the Sobolev spaces with $C^k(\mathbf{R}^d)$ in virtue of the Sobolev embedding Theorem. From Theorem 5.9 it follows in particular:

Corollary 5.10. *The fundamental solution of equation (5.19)*

$$K(t, x, y, \omega)$$

is well defined, as the unique solution of problem (5.19) with $u_0(x) = \delta_y(x)$, for all fixed $y \in \mathbf{R}^d$. We have

$$K \in C(\mathbf{R}_x^d; C_\mathcal{F}(\varepsilon, T; C^m(\mathbf{R}^d))$$

for all $\varepsilon \in (0, T)$ and $m \in \mathbf{N}$.

Here we meant that the function $x \mapsto K(t, x, y, \omega)$ is continuous from $\mathbf{R}^d$ to $C_\mathcal{F}(\varepsilon, T; C^m(\mathbf{R}^d))$.

One can show that, if u is the solution of equation (5.19) corresponding to the initial value u_0, then

$$u(t, x, \omega) = \int_{\mathbf{R}^d} K(t, x, y, \omega) u_0(y) dy,$$

whence one can derive the existence of a continuous version of the mapping $u_0 \mapsto u(t)$ (the stochastic flow). See [30] for details, where these results are essentially contained. They can also be extended to differential operators $\mathcal{A}$ and B^j of order $2m$ and (less or equal to) m of the form

$$\mathcal{A}u(x) = \sum_{|\alpha| \leq 2m} a_\alpha(x) D^\alpha u(x), \quad B^j u(x) = \sum_{|\beta| \leq m} b^j_\beta(x) D^\beta u(x) \tag{5.25}$$

under the assumption

$$\sum_{|\alpha|=2m} \left(a_\alpha(x) - \frac{1}{2} \sum_{j=1}^{n} \sum_{\beta_1+\beta_2=\alpha} b^j_{\beta_1}(x) b^j_{\beta_2}(x) \right) \xi^\alpha \neq 0$$

for all $(\xi_1, \ldots, \xi_d) \in \mathbf{R}^d - \{0\}$ and $x \in \mathbf{R}^d$. A carefull analysis of higher order problems in bounded and unbounded domains will be the object of a subsequent paper.

Remark — As far as the well posedness in $L^2(\mathbf{R}^d)$ is concerned, we can take $n = \infty$ without changing any assumption. The restriction to a finite number of operators B^j is only dictated by reasons of simplicity, when the regularity theory is concerned. For infinitely many B^j one has to impose complex restrictions on the coefficients of the differential operators in order to fulfill conditions (5.10) and (5.15) (notice the form (5.24) of the operators L^k). This remark applies also to the examples in bounded domains discussed in the next subsections.

5.3 Application to Second Order Equations in Bounded Domains: Dirichlet Boundary Condition

This and section 5.5 contain the main application of this paper. Let $D \subset R^d$ be a bounded open domain with regular boundary Γ. Consider the following stochastic parabolic equation in $[0, T] \times D$:

$$\begin{cases} du(t, x) = \mathcal{A}u(t, x)\, dt + \sum_{j=1}^{n} B^j u(t, x)\, dw^j(t), \\ u = 0, \quad t \in [0, T], x \in \Gamma, \\ u(0, x) = u_0(x), \quad x \in D, \end{cases} \tag{5.26}$$

where $\mathcal{A}$ and B_j are the differential operators of the previous section (see (5.20) (5.21)), with C^∞ coefficients in the closure of D, satisfying the standing joint ellipticity assumption (5.22).

In the following subsections we discuss separately the well posedness in Sobolev spaces with different orders, and we summarize the results in subsection 5.3.7. Now, a few general preliminary comments may be useful.

Remark 1 — In this application with Dirichlet boundary conditions we take $H = L^2(D)$, $D(A) = \{u \in H^2(D); u = 0 \text{ on } \Gamma\}$, and $Af = \mathcal{A}f$ for all $f \in D(A)$. Then for instance, $V_\alpha = H^\alpha(D)$ for $0 < \alpha < \frac{1}{2}$, $V_\alpha = \{u \in H^\alpha(D); u = 0 \text{ on } \Gamma\}$ for $\frac{1}{2} < \alpha < 2$. We do not write down explicitly the form of the boundary conditions in V_α for all α, but we just note that in the sequel we shall only use the following two facts: a) V_α is a closed subspace of $H^\alpha(D)$, with equivalent canonical norms; b) $V_{\alpha+2} = \{u \in H^{\alpha+2}(D); Au \in V_\alpha\}$ for all $\alpha \geq 0$. The operator A generates an analytic semigroup of contractions in H, which is also of negative type after a suitable translation, if necessary (see section 5.2).

Remark 2 — In the sequel we shall also be concerned with the adjoint of A. Here we have $D(A^*) = D(A)$,

$$A^*u(x) = \sum_{i,j=1}^{d} \frac{\partial}{\partial x_i}\left(a_{ij}(x)\frac{\partial u(x)}{\partial x_j}\right)$$

$$-\sum_{i=1}^{d} a_i(x)\frac{\partial u(x)}{\partial x_i} + \left(a(x) - \sum_{i=1}^{d}\frac{\partial a_i(x)}{\partial x_i}\right)u(x)$$

so that the spaces $V_\alpha^* = D((-A^*)^{\alpha/2})$ have the same properties of V_α (but in general they are different spaces for $|\alpha| > 2 + \frac{1}{2}$).

Remark 3 — The strategy to prove the well posedness and coercivity in the different spaces will be: i) to prove directly the coercivity condition (5.5) in $H = L^2(D)$ and $V_1 = H_0^1(D)$; ii) to apply the "shift" Theorems 5.6 and 5.8, in order to cover all the others V_s, $s \in \mathbf{Z}$, iii) to apply the interpolation Theorem 5.4 for the general V_α, $\alpha \in \mathbf{R}$. The coercivity in $L^2(D)$ is trivial and well known; in $H_0^1(D)$ it is difficult and seems to be new. In the application of the shift Theorems we have only to care about the assumptions of the form $B^j \in L(V_{\alpha+1}, V_\alpha)$. Indeed, consider for instance Theorem 5.6. Assumption i) has of course been previously proved; assumption ii) is the only one that needs a new analysis; assumption iii) is trivially fulfilled because for $u \in V_{\alpha+3}$, $L^j u \in V_\alpha$ by definition of L^j (we do not have to care about the boundary conditions in V_α), and the inequality (5.10) follows from the fact that the canonical norms in V_α and $H^\alpha(D)$ are equivalent, and the operators L^j are second order differential operators (the third order parts of AB^j and $B^j A$ coincide).

5.3.1 $L^2(D)$-Solutions

The well-posedness in $L^2(D)$ is well known. We repeat the simple basic computations for later reference.

Let us show that (5.5) is satisfied in $L^2(D)$ (with a different η). By (5.23) and integration by parts we have, for $u \in D(A)$,

$$\frac{1}{2}\sum_{k=1}^{n}|B^k u|_{L^2(D)}^2 = \frac{1}{2}\sum_{k=1}^{n}\int_D \sum_{i,j=1}^{d} b_i^k(x)b_j^k(x)\frac{\partial u(x)}{\partial x_i}\frac{\partial u(x)}{\partial x_j}\,dx$$

$$+\sum_{k=1}^{n}\int_D \sum_{i=1}^{d} b_i^k(x)\frac{\partial u(x)}{\partial x_i}c^k(x)u(x)\,dx + \frac{1}{2}\sum_{k=1}^{n}\int_D c^k(x)^2 u(x)^2\,dx$$

$$\leq \eta\int_D \sum_{i,j=1}^{d} a_{ij}(x)\frac{\partial u(x)}{\partial x_i}\frac{\partial u(x)}{\partial x_j}\,dx$$

$$+\rho\varepsilon\int_D \sum_{i=1}^{d}\left(\frac{\partial u(x)}{\partial x_i}\right)^2 dx + C_\varepsilon|u|_{L^2(D)}^2$$

$$\leq (\eta + \varepsilon) \int_D \sum_{i,j=1}^d a_{ij}(x) \frac{\partial u(x)}{\partial x_i} \frac{\partial u(x)}{\partial x_j} \, dx + C_\varepsilon |u|^2_{L^2(D)}$$

$$= -(\eta + \varepsilon) < Au, u >_{L^2(D)} + (\eta + \varepsilon) \int_\Gamma \frac{\partial u(x)}{\partial v_A} u(x) \, dx$$

$$+ (\eta + \varepsilon) < Cu, u >_{L^2(D)} + C_\varepsilon |u|^2_{L^2(D)} \tag{5.27}$$

for all $\varepsilon > 0$, and a suitable constant $C_\varepsilon > 0$. Here

$$\frac{\partial u(x)}{\partial v_A} = \sum_{i,j=1}^d a_{ij}(x) \frac{\partial u(x)}{\partial x_i} v_j(x)$$

where $v = (v_1, \ldots, v_d)$ is the outward normal to Γ, and C is the first order part of the differential operator $\mathcal{A}$:

$$Cu(x) = \sum_{i=1}^d a_i(x) \frac{\partial u(x)}{\partial x_i} + a(x)u(x) \tag{5.28}$$

Since $u \in D(A), u = 0$ on Γ. Thus (5.27) proves (5.5). In conclusion, we have the following classical result:

Theorem 5.11 *Equation (5.26) is well posed and coercive (see definition 5.3) in* $L^2(D)$.

5.3.2 $H^1(D)$-Solutions

Much more difficult is to check the coercivity condition (5.5) in $V_1 = H^1_0(D)$. The results of this section seem to be new.

A direct computation with the canonical norm in $H^1_0(D)$ leads to difficulties that we cannot overcome. It seems that the other natural norm

$$\int_D \left(u^2 + \sum_{i,j=1}^d a_{ij} \frac{\partial u}{\partial x_i} \frac{\partial u}{\partial x_j} \right) dx$$

allows us to prove (5.5), but some steps are particularly complex, and the argument do not generalize to the Neumann boundary condition of section 5.5. For these reasons we shall use an unusual but suitable norm in the next Theorem.

Theorem 5.12 *There exists a Hilbert topology in* $H^1(D)$ *(norm* $||.||_1$, *inner product* $< .,. >_1$*), equivalent to the canonical one, such that*

$$\frac{1}{2} \sum_{k=1}^{n} ||B^k u||_1^2 \leq -\eta < Au, u >_1 + \lambda ||u||_1^2 \tag{5.29}$$

for all $u \in V_3 = \{u \in H^3(D) : u = Au = 0 \text{ on } \Gamma\}$, *and for some constants* $\eta \in (0, 1)$ *and* $\lambda \geq 0$.

Proof. Step 1 — Define

$$||u||_1^2 = \int_D \left(u^2 + \sum_{i=1}^{N} |T_i u|^2 \right) dx \tag{5.30}$$

where the operators T_i are first order differential operators with C^∞ coefficients in the closure of D, with $T_1, \ldots, T_{N-1}$ tangential to Γ, and $T_N u = \frac{\partial u}{\partial v_{A^*}}$, where

$$\frac{\partial u}{\partial v_{A^*}} = \sum_{i,j=1}^{d} v_j a_{ji} \frac{\partial u}{\partial x_i}$$

(extended in the interior in any C^∞ way). We take these operators in such a way that (5.30) defines a norm in $H^1(D)$ equivalent to the canonical one.

The existence of operators with such properties can be shown by the following argument. Take two sequences of balls in $\mathbf{R}^d$, $U_1^0, \ldots, U_R^0$ and $U_1, \ldots, U_R$, with the following properties:

a) U_j^0 strictly contained in U_j;

b)   there exist C^∞ functions ϕ_j with compact support in U_j, $0 \leq \phi_j(x) \leq 1$, and $\phi_j(x) = 1$ in U_j^0;

c) $D \subset \cup U_j^0$;

d) $U_j \cap \overline{D}$ is diffeomorphic either to the unitary ball of R^d or to the unitary semiball of $R_+^d = \{x = (x_1, \ldots, x_d) \in R^d : x_d \geq 0\}$, with $U_j \cap \Gamma$ sent in $\{x \in R^d : x_d = 0, |x| \leq 1\}$.

Over each $U_j \cap \overline{D}$ one can define (by (d)) differential operators $T_1^j, \ldots, T_{d-1}^j$, tangential to Γ, such that the norm

$$\int_{U_j^0 \cap D} \left(u^2 + \sum_{i=1}^{d-1} |T_i u|^2 + \frac{\partial u}{\partial v_{A^*}} \right) dx$$

is equivalent to the canonical one. The operators $\phi_j T_i^j$ have compact support in $U_j \cap D$ and can be extended to 0 over the other part of D, being globally tangential to Γ (but degenerate). Let us denote by $T_1, \ldots, T_{N-1}$ the operators $\{\phi_j T_i^j ; j = 1, \ldots R, i = 1, \ldots, d-1\}$. By the property (c) and the fact that $\phi_j(x) = 1$ in U_j^0, one can check that (5.30) is equivalent to the standard topology of $H^1(D)$.

Step 2 — Since $T_i B^k - B^k T_i$ is a first order differential operator, for all $\varepsilon > 0$ we have

$$\frac{1}{2} \sum_{k=1}^n \|B^k u\|_1^2 = \frac{1}{2} \sum_{k=1}^n \int_D \left(|B^k u|^2 + \sum_{i=1}^N |T_i B^k u|^2 \right) dx$$

$$\leq \frac{1+\varepsilon}{2} \sum_{k=1}^n \int_D \sum_{i=1}^N |B^k T_i u|^2 \, dx + r_1(u) + (1 + \frac{1}{\varepsilon}) r_2(u),$$

where $r_1(u)$ and $r_2(u)$ are terms bounded by a constant times $\|u\|_1^2$. Hence, using the standing inequality (5.23), we can find two constants $\eta' \in (0, 1)$ and $\lambda' \geq 0$ such that

$$\frac{1}{2} \sum_{k=1}^n \|B^k u\|_1^2 \leq \eta' \int_D \sum_{i=1}^N \sum_{j,l=1}^d a_{jl} \frac{\partial T_i u}{\partial x_j} \frac{\partial T_i u}{\partial x_l} \, dx + \lambda' \|u\|_1^2.$$

Therefore there exist constants $\eta \in (0, 1)$ and $\delta > 0$ such that (by Green formula)

$$\frac{1}{2} \sum_{k=1}^n \|B^k u\|_1^2 \leq -\delta \int_D \sum_{i=1}^N \sum_{j,l=1}^d a_{jl} \frac{\partial T_i u}{\partial x_j} \frac{\partial T_i u}{\partial x_l} \, dx$$

$$- \eta \int_D \sum_{i=1}^N \sum_{j,l=1}^d \frac{\partial}{\partial x_l} \left(a_{jl} \frac{\partial T_i u}{\partial x_j} \right) T_i u \, dx$$

$$+ \eta \int_\Gamma \sum_{i=1}^N \sum_{j,l=1}^d \nu_l a_{jl} \frac{\partial T_i u}{\partial x_j} T_i u \, d\sigma + \lambda' \|u\|_1^2.$$

On one side,

$$\int_D \sum_{i=1}^N \sum_{j,l=1}^d a_{jl} \frac{\partial T_i u}{\partial x_j} \frac{\partial T_i u}{\partial x_l} \, dx \geq c_0 \int_D \sum_{i=1}^N \sum_{j=1}^d |\frac{\partial T_i u}{\partial x_j}|^2 \, dx$$

$$\geq c_1 \|u\|_{H^2(D)}^2 - c_2 \|u\|_{L^2(D)}^2,$$

by the strong ellipticity of a_{ij} (cf. (5.22)). On the other side, it is strightforward to check that

$$\int_D \sum_{i=1}^N \sum_{j,l=1}^d \frac{\partial}{\partial x_l} \left(a_{jl} \frac{\partial T_i u}{\partial x_j} \right) T_i u \, dx = < Au, u >_1 + \rho(u),$$

where

$$|\rho(u)| \leq c\|u\|_{H^1(D)}\|u\|_{H^2(D)} \leq \varepsilon\|u\|_{H^2(D)}^2 + \frac{1}{4\varepsilon}\|u\|_{H^1(D)}^2$$

for all $\varepsilon > 0$. Collecting these inequalities and choosing ε sufficiently small, we can find two constants $\delta' > 0$ and $\lambda'' \geq 0$ such that

$$\frac{1}{2}\sum_{k=1}^{n}\|B^k u\|_1^2 \leq -\delta'\|u\|_{H^2(D)}^2 - \eta < Au, u >_1 +\lambda''\|u\|_1^2$$

$$+\eta\int_{\Gamma}\sum_{i=1}^{N}\sum_{j,l=1}^{d} v_l a_{jl}\frac{\partial T_i u}{\partial x_j}T_i u \, d\sigma.$$

The proof of the Theorem is complete if we show that for all $\varepsilon > 0$ there exists a constant $c(\varepsilon) > 0$ such that

$$\int_{\Gamma}\sum_{i=1}^{N}\sum_{j,l=1}^{d} v_l a_{jl}\frac{\partial T_i u}{\partial x_j}T_i u \, d\sigma \leq \varepsilon\|u\|_{H^2(D)}^2 + c(\varepsilon)\|u\|_{H^1(D)}^2.$$

Step 3 — Here we use for the first time the assumption that $u \in V_3$. Since $u = 0$ on Γ, also $T_i u = 0$ on Γ for $j < N$ (the operators T_i are tangential to the boundary). Thus we have

$$\int_{\Gamma}\sum_{i=1}^{N}\sum_{j,l=1}^{d} v_l a_{jl}\frac{\partial T_i u}{\partial x_j}T_i u \, d\sigma = \int_{\Gamma}\left(\frac{\partial}{\partial v_A}\frac{\partial u}{\partial v_{A^*}}\right)\frac{\partial u}{\partial v_{A^*}} \, d\sigma.$$

Denoting by a the matrix $\{a_{ij}\}$, and by $(.|.)$ the scalar product in $\mathbf{R}^d$, we want to show that, on Γ,

$$\frac{\partial}{\partial v_A}\frac{\partial u}{\partial v_{A^*}} = (av|v)Au + R_1 u$$

for some first order differential operator R_1. By mapping a local neighborhood of Γ into a semi-ball (as in step 1), we see that it is sufficient to prove a similar relation in the case of a set D of the form $D = \{x = (x_1, \ldots, x_d) \in \mathbf{R}^d : x_d \geq 0\}$, with Γ given by $\{x \in \mathbf{R}^d : x_d = 0\}$ (indeed, note that under a diffeomorphism the previous relation transforms itself into a relation of the same form with the new operator A). In this case, $v = e_n$, the last vector a the canonical basis, and $(av|v) = a_{nn}$. Moreover, *up to first order differential operators*, we have

$$\frac{\partial}{\partial v_A}\frac{\partial u}{\partial v_{A^*}} = \sum_{j,l=1}^{d} a_{ni}a_{jn}\frac{\partial}{\partial x_i}\frac{\partial u}{\partial x_j}$$

$$= a_{nn}\{a_{nn}\frac{\partial^2 u}{\partial x_n^2} + \sum_{i<n} a_{ni}\frac{\partial^2 u}{\partial x_i \partial x_n} + \sum_{j<n} a_{jn}\frac{\partial^2 u}{\partial x_n \partial x_j}\}$$

and

$$(av|v)Au = a_{nn}\sum_{i,j=1}^{d} a_{ij}\frac{\partial^2 u}{\partial x_i \partial x_j}$$

$$== a_{nn}\{a_{nn}\frac{\partial^2 u}{\partial x_n^2} + \sum_{i<n} a_{ni}\frac{\partial^2 u}{\partial x_i \partial x_n} + \sum_{j<n} a_{jn}\frac{\partial^2 u}{\partial x_n \partial x_j}\}$$

where we have used the fact that $u = 0$ on the boundary, so that all second derivatives in variables both different from x_n vanish. We have proved the desired relation.

Since $Au = 0$ on Γ (by the assumption that $u \in V_3$), we finally see that

$$\int_{\Gamma} \sum_{i=1}^{N} \sum_{j,l=1}^{d} v_l a_{jl}\frac{\partial T_i u}{\partial x_j} T_i u \, d\sigma = \int_{\Gamma} R_1 u \frac{\partial u}{\partial v_{A^*}} \, d\sigma.$$

Recall that the trace γ on Γ is a linear bounded operator from $H^{\frac{1}{2}+\delta}(D)$ into $H^{\delta}(D)$, for all $\delta > 0$. Therefore, both γR_1 and $\frac{\partial}{\partial v_{A^*}}$ are bounded linear operators from $H^{\frac{3}{2}+\delta}(D)$ into $H^{\delta}(D)$, for all $\delta > 0$. Taking any $\delta \in (0, \frac{1}{2})$, it follows that

$$\int_{\Gamma} R_1 u \frac{\partial u}{\partial v_{A^*}} \, d\sigma \le c(\delta)\|u\|^2_{H^{\frac{3}{2}+\delta}(D)} \le \varepsilon\|u\|^2_{H^2(D)} + c(\varepsilon)\|u\|^2_{H^1(D)},$$

for all $\varepsilon > 0$, by a well known interpolation inequality. This completes the proof of the Theorem.

In order to apply Theorem 5.2 in $V_1 = H_0^1(D)$ the operators B^j have to map V_2 in V_1. This is satisfied if and only if

$$b^j(x) \quad \text{tangent to} \quad \Gamma, \tag{5.31}$$

where $b^j(x) = (b_1^j(x), \ldots, b_d^j(x))$ (see also (4.13) and section 4). Indeed, if $u \in V_2$, ∇u is orthogonal to Γ (since $u = 0$ on Γ). Therefore $B^j u = 0$ on Γ, i.e. the boundary condition in the requirement $B^j u \in V_1$ is fulfilled (the converse is similarly true). Of course, also the regularity condition $B^j u \in H^1(D)$ is satisfied. From Theorem 5.2 we finally have:

Theorem 5.13 *Under the assumption (5.31), equation (5.26) is well posed and coercive in $V_1 = H_0^1(D)$.*

5.3.3 $H^2(D)$-Solutions

Having proved the well posedness, and in particular the coercivity condition (5.5), in $L^2(D)$, we can use the "shift" Theorem 5.6 to obtain the well-posedness in $V_2 = D(A)$ (which has being roughly indicated by $H^2(D)$ in the title). To this end recall Remark 3 of section 5.3. Condition ii) of Lemma 5.5 on B^j is satisfied under the assumption (5.31), exactly as in the previous subsection. Hence we have:

Theorem 5.14 *Under the assumption (5.31), equation (5.26) is well posed and coercive in $V_2 = D(A)$.*

5.3.4 $H^n(D)$-Solutions, $n \geq 3$

Using iteratively Theorem 5.6, and the coercivity conditions in H and V_1 just proved, one can obtain analogous results in V_n for all $n \geq 3$. The only point to be discussed in each case is the condition

$$B^j : V_{n+1} \to V_n. \tag{5.32}$$

For $n \geq 3$ we do not have a characterization of (5.32) in terms of the coefficients of A and B^j, but only sufficient conditions. A simple (but certainly not optimal) one is given in the following Theorem.

Theorem 5.15 *If we assume*

$$b_l^k = \frac{\partial b_l^k}{\partial x_i} = \frac{\partial^2 b_l^k}{\partial x_i \partial x_j} = 0, \quad c^k = \frac{\partial c^k}{\partial x_i} = 0 \quad \text{on} \ \Gamma, \tag{5.33}$$

for all the values of k, l, i, j, then condition (5.32) is satisfied, and thus problem (5.26) is well posed, in both $V_3 = \{u \in H^3(D) : u = Au = 0$ on $\Gamma\}$ and $V_4 = \{u \in H^4(D) : u = Au = 0$ on $\Gamma\}$.

In general, if we assume

$$D^\alpha b_l^k = 0, \quad D^\beta c^k = 0 \quad \text{on} \ \Gamma, \tag{5.34}$$

for all multi-indexes α and β with $|\alpha| \leq 2n$ and $|\beta| \leq 2n - 1$, and for all the values of k, l, then condition (5.32) is satisfied, and thus problem (5.26) is well posed, in both V_{2n+1} and V_{2n+2}.

Proof — Consider the statement relative to V_3. We have to prove that

$$B^j u = 0 \tag{5.35}$$

and

$$AB^j u = 0 \tag{5.36}$$

on Γ, for all $u \in V_4$. Of course (5.35) holds true since assumption (5.33) implies (5.31), which in turns was sufficient for (5.35). On the other side,

$$AB^j u = A(b^j \nabla u) + (Ac^j)u + R^j u, \tag{5.37}$$

where the second order differential operator R^j contains only derivatives of c^j up to order 1. Recalling that $u = 0$ on Γ (for the second term of the right-hand-side of (5.37)), it is now easy to see that (5.36) is fulfilled under the assumption (5.33). The proof of the other statements is similar.

5.3.5 $H^{-1}(D)$-Solutions

Here we use Theorem 5.8 starting from the well posedness in V_1. We have to check the condition $B^j(L^2(D)) \subset H^{-1}(D)$. This holds true without restrictions on the coefficients of B^j since $(B^j)^*(H_0^1(D)) \subset L^2(D)$ (see Lemma 5.17 below). Hence we are led to think that problem (5.26) is well posed in $H^{-1}(D)$ without restrictions on the coefficients of B^j. Unfortunately, we cannot prove this plausible results by means of Theorem 5.8, because the well posedness in $H_0^1(D)$ is satisfied only under the condition (5.31). One should prove the coercivity condition (5.5) in $H^{-1}(D)$ directly, without the abstract Theorem 5.8, but such direct proof is not easy (if true). Thus, the result of this section seems to be not optimal.

By the previous arguments we at least have:

Theorem 5.16 *Under the assumption (5.31), equation (5.26) is well posed and coercive in $V_{-1} = H^{-1}(D)$.*

5.3.6 $H^{-n}(D)$-Solutions, $n \geq 2$

From the point of view of the restrictions on b_i^j and c^j, the results are (roughly) symmetric with respect to $V_{-1/2}$, not with respect to H. The reason is simply that the assumption

$$B^j \in L(V_{-\alpha+1}, V_{-\alpha}) \tag{5.38}$$

is equivalent to

$$(B^j)^* \in L(V_\alpha^*, V_{\alpha-1}^*), \tag{5.39}$$

where $(B^j)^*$ is the adjoint of B^j with respect to H. Here we denote by V_α^* the spaces $D((-A^*)^{\alpha/2})$, defined in terms of A^* (adjoint of A with respect to H), similarly to the spaces V_α of section 5.1.

Lemma 5.17 *Let $\alpha > 0$ be given. Let $(B^j)^*$ be the differential operator defined as*

$$(B^j)^* u = -\sum_{i=1}^{d} b_i^j \frac{\partial u}{\partial x_i} + c^j u - (\operatorname{div} b^j)u. \tag{5.40}$$

Then (5.38) and (5.39) are equivalent.

Proof — We only prove that (5.39) implies (5.38) (the statement that will be used in the sequel), the converse being analogous. As a preliminary, we note that, by the Green formula,

$$< B^j u, v >_{L^2(D)} = \int_D u \left[-\sum_{i=1}^d b_i^j \frac{\partial v}{\partial x_i} - (\text{ div } b^j)v + c^j v \right] dx$$

$$+ \int_\Gamma \sum_{i=1}^d b_i^j \nu_i u v \, d\sigma \tag{5.41}$$

for all $u, v \in H^1(D)$.

Assume that (5.39) holds (α is fixed) for the operator defined by (5.40). Let $((B^j)^*)'$ be the dual operator of $(B^j)^*$, which is, by definition, a bounded linear operator from $V_{-\alpha+1}$ to $V_{-\alpha}$ (it is strightforward to verify that V_β^* and $V_{-\beta}$ are dual spaces for all β, with respect to H, after proper identifications). It is sufficient to prove that $((B^j)^*)'$ coincides with B^j. Since

$$< ((B^j)^*)' u, v >_{V_{-\alpha}, V_\alpha^*} = < u, (B^j)^* v >_{V_{-\alpha+1}, V_{\alpha-1}^*}$$

$$= < u, (B^j)^* v >_{L^2(D)} = < B^j u, v >_{L^2(D)}$$

for all $u \in V_1$ and $v \in V_\alpha^* \cap V_1^*$ (cf. (5.41) for the last identity), by the density of V_α in $L^2(D)$ we have $((B^j)^*)' u \in L^2(D)$ and $((B^j)^*)' u = B^j u$, for all $u \in V_1$. This proves that $((B^j)^*)'$ coincides with B^j (since for $\alpha < 1$ the differential operator $B^j : V_\alpha \to H^{\alpha-1}(D)$ is understood as the unique extension of $B^j : V_1 \to L^2(D)$). The proof is complete.

For $n = 2, 3$ we have:

Theorem 5.18 *Under the assumption (5.31), equation (5.26) is well posed in V_{-2} and V_{-3}.*

Proof — For $n = 2$ the property $B^j \in L(V_{-1}, V_{-2})$ follows from

$$(B^j)^* \in L(V_2, V_1). \tag{5.42}$$

The latter holds under (5.31) as in the previous sections. For $n = 3$ the property $B^j \in L(V_{-2}, V_{-3})$ follows from $(B^j)^* \in L(V_3, V_2)$ as in the previous case.

In general we have:

Theorem 5.19 *Under the assumption (5.34), equation (5.26) is well posed in V_{-2n-2} and V_{-2n-3}.*

5.3.7 Summary of Regularity Results for Dirichlet Boundary Value Problem. $H^\alpha(D)$-Solutions, $\alpha \in \mathbf{R}$

Using the interpolation Theorem 5.4, one can obtain well-posedness results in V_{s+r}, $s \in \mathbf{Z}$, $r \in (0, 1)$, assuming the well-posedness in V_s and V_{s+1}. We remark that they are not optimal, since the conditions on b_i^k and c^k imposed to have the well-posedness in V_s and V_{s+1} are sometimes redundant with respect to the requirement $B^k \in L(V_{s+r+1}, V_{s+r})$. For instance, $V_\alpha = H^\alpha(D)$ for $\alpha \in [0, \frac{1}{2})$ (no boundary conditions), so that the well posedness in such V_α should not require any condition on b_i^k, in contrast to the well-posedness in $V_1 = H_0^1(D)$. One should check explicitly the coercivity condition in V_{s+r}, without using the interpolation Theorem 5.4, but we do not know how to perform this computation.

Theorem 5.20 *Consider equation (5.26), with A and B^j defined by (5.20)–(5.21) with C^∞-coefficients in $\overline{D}$, and satisfying the joint ellipticity condition (5.22). Then:*

i) equation (5.26) is well posed and coercive in $H = L^2(D)$;

ii) if the vector fields b^j are tangent to Γ, then equation (5.26) is well posed in V_α for all $-3 \le \alpha \le 2$, and the coercivity condition (5.5) is satisfied in such V_α with integer α;

iii) if

$$b_l^k = \frac{\partial b_l^k}{\partial x_i} = \frac{\partial^2 b_l^k}{\partial x_i \partial x_j} = 0, \quad c^k = \frac{\partial c^k}{\partial x_i} = 0 \quad \text{on } \Gamma,$$

for all the values of k, l, i, j, then equation (5.26) is well posed in V_α for all $-5 \le \alpha \le 4$, and the coercivity condition (5.5) is satisfied in such V_α with integer α;

iv) in general, if we assume

$$D^\alpha b_l^k = 0, \quad D^\beta c^k = 0 \quad \text{on } \Gamma,$$

for all multi-indexes α and β with $|\alpha| \le 2n$ and $|\beta| \le 2n - 1$, and for all the values of k, l, then equation (5.26) is well posed in V_α for all $-(2n+3) \le \alpha \le 2n+2$, and the coercivity condition (5.5) is satisfied in such V_α with integer α;

v) finally, if the coefficients b_i^k and c^k have compact support in D, then equation (5.26) is well posed in every V_α, $\alpha \in \mathbf{R}$, and coercive if $\alpha \in \mathbf{Z}$.

5.3.8 Regularity for the Adjoint Equation

For later use in section 7, it is convenient to state now some regularity results for the equation

$$du = A^*u\, dt + \sum_{j=1}^n (B^j)^*u\, dw^j, \quad u(0) = u_0, \tag{5.43}$$

which is called here *adjoint equation* to (5.26). The operator A^* is the adjoint of A in H, ad described in Remark 2 of section 5.3, while $(B^j)^*$ have been defined and discussed in section 5.3.6. Arguing as above, we have the following Theorem, which summarizes the various cases.

Theorem 5.21 *i) equation (5.43) is well posed and coercive in $H = L^2(D)$;*

ii) if the vector fields b^j are tangent to Γ, then equation (5.43) is well posed in V_α^* for all $-3 \leq \alpha \leq 2$;

iii) if

$$b_l^k = \frac{\partial b_l^k}{\partial x_i} = \frac{\partial^2 b_l^k}{\partial x_i \partial x_j} = 0, \quad c^k = \frac{\partial c^k}{\partial x_i} = 0 \quad \text{on} \quad \Gamma,$$

for all the values of k, l, i, j, then equation (5.43) is well posed in V_α^* for all $-5 \leq \alpha \leq 4$;

iv) in general, if we assume

$$D^\alpha b_l^k = 0, \quad D^\beta c^k = 0 \quad \text{on} \quad \Gamma,$$

for all multi-indexes α and β with $|\alpha| \leq 2n$ and $|\beta| \leq 2n - 1$, and for all the values of k, l, then equation (5.43) is well posed in V_α^* for all $-(2n + 3) \leq \alpha \leq 2n + 2$;

v) finally, if the coefficients b_i^k and c^k have compact support in D, then equation (5.43) is well posed in every V_α^*, $\alpha \in \mathbf{R}$.

5.4 Application to Equations of Order $2m$ in Bounded Domains: Dirichlet Boundary Conditions

Just as an example of application to parabolic problems of order greater then 2, consider the problem

$$\begin{cases} du(t, x) = \mathcal{A}u(t, x) \, dt + \sum_{j=1}^n B^j u(t, x) \, dw^j(t), \\ u = 0, \frac{\partial u}{\partial v} = 0, \ldots, \frac{\partial^{m-1} u}{\partial v^{m-1}} = 0 \quad t \in [0, T], x \in \Gamma, \\ u(0, x) = u_0(x), \quad x \in D, \end{cases} \tag{5.44}$$

where $\mathcal{A}$ and B^j are the differential operators given by (5.25), and $\frac{\partial u}{\partial v}, \ldots, \frac{\partial^{m-1} u}{\partial v^{m-1}}$ are the successive normal derivatives on the boundary. Due to the simple boundary conditions it is not difficult to repeat the computation (5.27) in $L^2(D)$. Then assuming that

$$b_\beta^j \quad \text{have compact support in} \quad D, \tag{5.45}$$

we can proceed as in section 5.3, to get:

Theorem 5.22 *Under the assumption (5.45), equation (5.44) is well posed in all spaces* V_α, $\alpha \in \mathbf{R}$, *and coercive for* $\alpha \in \mathbf{Z}$.

5.5 Application to Second Order Equations in Bounded Domains: Neumann Boundary Condition

Consider now the Neumann boundary value problem analogous to (5.26):

$$\begin{cases} du(t, x) = \mathcal{A}u(t, x)\, dt + \sum_{j=1}^{n} B^j u(t, x)\, dw^j(t), \\ \frac{\partial u(t,x)}{\partial \nu_A} = 0, \quad t \in [0, T], x \in \Gamma, \\ u(0, x) = u_0(x), \quad x \in D, \end{cases} \tag{5.46}$$

where $\mathcal{A}$ and B^j are given by (5.20)–(5.21), satisfying the joint ellipticity condition (5.22). In this case, we take $H = L^2(D)$, $D(A) = \{u \in H^2(D); \frac{\partial u}{\partial \nu_A} = 0$ on $\Gamma\}$, and $Af = \mathcal{A}f$ for all $f \in D(A)$. Then for instance, $V_\alpha = H^\alpha(D)$ for $0 < \alpha < \frac{3}{2}$, $V_\alpha = \{u \in H^\alpha(D); \frac{\partial u}{\partial \nu_A} = 0$ on $\Gamma\}$ for $\frac{3}{2} < \alpha < 2$. Properties a) and b) described in Remark 1 of section 5.3 holds true also here. The operator A generates an analytic semigroup of contractions in H, which is also of negative type after a suitable translation, if necessary.

Remark 1 — The adjoint operators are slightly more complex than in the Dirichlet case (this will have some consequences on the complexity and non-symmetry of the compatibility conditions on b_i^k and c^k). Here we have (by Green formula)

$$D(A^*) = \{u \in H^2(D); \frac{\partial u}{\partial \nu_{A^*}} - (\sum_{i=1}^{d} a_i \nu_i)u = 0 \text{ on } \Gamma\} \tag{5.47}$$

which reduces to the more familiar one

$$D(A^*) = \{u \in H^2(D); \frac{\partial u}{\partial \nu_{A^*}} = 0 \text{ on } \Gamma\} \quad \text{if} \quad \sum_{i=1}^{d} a_i \nu_i = 0 \tag{5.48}$$

(for the definition of $\frac{\partial u}{\partial \nu_{A^*}}$ see section 5.3.2). The operator A^* is still given by

$$A^* u(x) = \sum_{i,j=1}^{d} \frac{\partial}{\partial x_i} \left(a_{ij}(x) \frac{\partial u(x)}{\partial x_j} \right)$$

$$- \sum_{i=1}^{d} a_i(x) \frac{\partial u(x)}{\partial x_i} + \left(a(x) - \sum_{i=1}^{d} \frac{\partial a_i(x)}{\partial x_i} \right) u(x)$$

The spaces $V_\alpha^* = D((-A^*)^{\alpha/2})$ have the same properties of V_α (in general these spaces are different even for for $|\alpha| > \frac{3}{2}$), but the expressions involving the boundary conditions of V_α^* will be less simple, in general.

As to the adjoint of B^j, we have the following Lemma. The proof, based on equation (5.41), is the same as that of Lemma 5.17.

Lemma 5.23 *Let $\alpha > 0$ be given. Assume that the vector fields b^j are tangent to Γ, i.e.*

$$\sum_{i=1}^{d} b_i^j \nu_i = 0. \tag{5.49}$$

Let $(B^j)^*$ be the differential operator defined as

$$(B^j)^* u = -\sum_{i=1}^{d} b_i^j \frac{\partial u}{\partial x_i} + c^j u - (\text{ div } b_j) u. \tag{5.50}$$

Then the two properties

$$B^j \in L(V_{-\alpha+1}, V_{-\alpha}) \tag{5.51}$$

and

$$(B^j)^* \in L(V_\alpha^*, V_{\alpha-1}^*), \tag{5.52}$$

are equivalent.

Remark 2 — The strategy to prove the well posedness and coercivity in the different spaces is the same used for the Dirichlet problem: i) we prove directly the coercivity condition (5.5) in $H = L^2(D)$ and $V_1 = H^1(D)$ (the latter by a slight modification of Theorem 5.12); ii) we apply the "shift" Theorems 5.6 and 5.8 for the other spaces V_α, $\alpha \in Z$, and the interpolation Theorem 5.4 for the general V_α, $\alpha \in R$. As before, in the application of the shift Theorems we have only to care about the assumptions of the form $B^j \in L(V_{\alpha+1}, V_\alpha)$.

5.5.1 Regularity Results for the Neumann Boundary Value Problem

Theorem 5.24 *Consider equation (5.46) with A and B^j defined by (5.20)–(5.21) (C^∞-coefficients in $\overline{D}$ satisfying the joint ellipticity condition (5.22)). Then:*

i) equation (5.46) is well posed and coercive in $H = L^2(D)$;

ii) equation (5.46) is well posed and coercive in $V_1 = H^1(D)$; hence, it is also well posed in $V_\alpha = H^\alpha(D)$ for $0 < \alpha < 1$;

iii) if the vector fields b^j are tangent to Γ, then equation (5.46) is well posed and coercive in V_{-2} and V_{-1}; hence it is also well posed in V_α, for $-2 \leq \alpha \leq 1$;

iv) if

$$b_i^j = \frac{\partial b_i^j}{\partial \nu_A} = \frac{\partial c^j}{\partial \nu_A} = 0 \quad \text{on } \Gamma, \tag{5.53}$$

for all i, j, then equation (5.46) is well posed and coercive in $V_2 = \{u \in H^2(D); \frac{\partial u}{\partial v_A} = 0$ on $\Gamma\}$ and in $V_3 = \{u \in H^3(D); \frac{\partial u}{\partial v_A} = 0$ on $\Gamma\}$; hence (since the assumption of part (iii) is satisfied) it is also well posed in V_α, for $-2 \leq \alpha \leq 3$;

v) if for all i, j, k

$$b_i^j = \frac{\partial b_i^j}{\partial x_k} = c^j = \frac{\partial c^j}{\partial x_k} = 0 \quad \text{on} \quad \Gamma, \tag{5.54}$$

and

$$\text{either} \quad \text{div } b^j = 0 \quad \text{in} \quad D, \quad \text{or} \quad \frac{\partial^2 b_i^j}{\partial x_k \partial x_l} = 0 \quad \text{on} \quad \Gamma, \tag{5.55}$$

then equation (5.46) is well posed and coercive in V_{-4} and V_{-3}; hence (since the assumptions of part (iv) are satisfied) it is also well posed in V_α, for $-4 \leq \alpha \leq 3$;

vi) if

$$D^\alpha b_i^j = D^\alpha c^j = 0 \quad \text{on} \quad \Gamma, \tag{5.56}$$

for all multi-indexes α such that $|\alpha| \leq 2n - 1$, then equation (5.46) is well posed in V_α, for $-2n \leq \alpha \leq 2n + 1$, and it is coercive in all such spaces with integer α;

vii) if, in addition to (5.56), we have

$$\text{either} \quad \text{div } b^j = 0 \quad \text{in} \quad D, \quad \text{or} \quad D^\alpha b_i^j = 0 \quad \text{on} \quad \Gamma \tag{5.57}$$

also for $|\alpha| = 2n$, then equation (5.46) is well posed in V_α, for $-2n - 2 \leq \alpha \leq 2n + 1$, and it is coercive in all such spaces with integer α;

viii) finally, if the coefficients b_i^k and c^k have compact support in D, then equation (5.46) is well posed in every V_α, $\alpha \in \mathbf{R}$, and coercive in every V_α, $\alpha \in \mathbf{Z}$.

Proof. Part i) — Exactly the same computation of section 5.3.1 leads to the coercivity condition in $L^2(D)$ for the Neumann problem, since $\frac{\partial u}{\partial v_A} = 0$ by $u \in D(A)$.

Part ii) — Since V_1 does not contain any boundary condition, we do not have to impose any restriction on the coefficients of B^j. The only point to prove is the coercivity condition (5.5) in V_1. To this end we have a complete analog of Theorem 5.12:

Lemma 5.25 *There exists a Hilbert topology in $H^1(D)$ (norm $\|.\|_1$, inner product $< ., . >_1$), equivalent to the canonical one, such that*

$$\frac{1}{2} \sum_{k=1}^{n} \|B^k u\|_1^2 \leq -\eta < Au, u >_1 + \lambda \|u\|_1^2 \tag{5.28}$$

for all $u \in V_3 = \{u \in H^3(D) : \frac{\partial u}{\partial v_A} = 0$ on $\Gamma\}$, and for some constants $\eta \in (0, 1)$ and $\lambda \geq 0$.

Proof of Lemma 5.25 — The first to steps of the proof of Theorem 5.12 are the same, except for a more convenient definition of T_N:

$$T_N u = \frac{\partial u}{\partial \nu_A}.$$

Let us come to step 3. Here we have

$$\int_\Gamma \sum_{i=1}^N \sum_{j,l=1}^d \nu_l a_{jl} \frac{\partial T_i u}{\partial x_j} T_i u \, d\sigma = \sum_{i=1}^N \int_\Gamma \frac{\partial T_i u}{\partial \nu_A} T_i u \, d\sigma$$

$$= \sum_{i=1}^N \int_\Gamma \left(T_i \frac{\partial u}{\partial \nu_A} \right) T_i u \, d\sigma + \sum_{i=1}^N \int_\Gamma R_i u T_i u \, d\sigma,$$

where $R_i u$ is a first order differential operator. Now, on Γ, $T_N u = 0$ and $T_i \frac{\partial u}{\partial \nu_A} = 0$ for $i < N$ because $\frac{\partial u}{\partial \nu_A} = 0$ and the operators T_i are tangential. Hence

$$\int_\Gamma \sum_{i=1}^N \sum_{j,l=1}^d \nu_l a_{jl} \frac{\partial T_i u}{\partial x_j} T_i u \, d\sigma = \sum_{i=1}^N \int_\Gamma R_i u T_i u \, d\sigma \leq c|u|^2_{H^{3/2}(D)}.$$

Then it is sufficient to repeat the last part of the proof of step 3 of Theorem 5.12. The proof is complete.

Let us continue the proof of Theorem 5.24. By the previous Lemma and the preceeding comments we have the well-posedness and coercivity in V_1, as claimed. The well-posedness in $V_\alpha = H^\alpha(D)$ for $0 < \alpha < 1$ follows then by interpolation (Theorem 5.4).

Part iii) — This follows from Theorem 5.8, shifting back from the well-posedness and coercivity in H and V_1. The assumption that the vector fields b^j are tangent to Γ is imposed to apply Lemma 5.23. It is easy to see that no other conditions are needed to invoke Theorem 5.8 (in particular, note that $V_1^* = H^1(D)$, without boundary conditions, so that $(B^j)^* \in L(V_2, V_1)$ without restrictions on the coefficients).

Part iv) — We apply Theorem 5.6 starting from the well-posedness and coercivity in H and V_1. The only conditions that we have to check are $B^j \in L(V_3, V_2)$ and $B^j \in L(V_4, V_3)$. In both cases we have to prove that $\frac{\partial B^j u}{\partial \nu_A} = 0$, for $u \in V_3$ or V_4 depending on the case (but this makes no difference). We have

$$\frac{\partial B^j u}{\partial \nu_A} = \frac{\partial}{\partial \nu_A} \left(\sum_{i=1}^d b_i^j \frac{\partial u}{\partial x_i} \right) + \frac{\partial c^j u}{\partial \nu_A}.$$

On one side,

$$\frac{\partial}{\partial \nu_A} \left(\sum_{i=1}^d b_i^j \frac{\partial u}{\partial x_i} \right) = \sum_{i=1}^d \frac{\partial b_i^j}{\partial \nu_A} \frac{\partial u}{\partial x_i} + \sum_{i=1}^d b_i^j \frac{\partial}{\partial \nu_A} \frac{\partial u}{\partial x_i} = 0$$

by the assumptions (5.53). On the other hand,

$$\frac{\partial c^j u}{\partial v_A} = u \frac{\partial c^j}{\partial v_A} + c^j \frac{\partial u}{\partial v_A};$$

the first term is equal to 0 by assumption (5.53), the second by the condition $\frac{\partial u}{\partial v_A} = 0$, included in $u \in V_3$. Thus $\frac{\partial B^j u}{\partial v_A} = 0$, and the proof of part iv) is complete.

Part v) — Recalling the expression for $(B^j)^*$, we see that (5.54)–(5.55) are simple sufficient conditions to have the vector fields b^j tangent to Γ (in order to apply Lemma 5.23) and

$$\left(\frac{\partial}{\partial v_{A^*}} - \sum_{i=1}^{d} a_i v_i \right) (B^j)^* u = 0. \tag{5.59}$$

This implies that $(B^j)^* \in L(V_4^*, V_3^*)$ and $(B^j)^* \in L(V_3^*, V_2^*)$ (as in part (iv)). Hence, by Lemma 5.23, $B^j \in L(V_{-3}, V_{-4})$ and $B^j \in L(V_{-2}, V_{-3})$, so that we can apply Theorem 5.8 to get the well-posedness in V_{-4} and V_{-4}. Then we conclude by interpolation.

Parts vi)–viii) — The proof of parts vi)-vii) is similar to iv)-v). Finally, part viii) is a corollary of the previous results.

5.5.2 Regularity for the Adjoint Equation

Consider the adjoint equation to (5.46)

$$du = A^* u \, dt + \sum_{j=1}^{n} (B^j)^* u \, dw^j, \quad u(0) = u_0, \tag{5.60}$$

The proof of the following Theorem is entirely similar to that of Theorem 5.24. The only minor difference is that in the proof of the coercivity conditions in $H = L^2(D)$ and $V_1^* = H^1(D)$ some lower order terms appear since $\frac{\partial u}{\partial v_{A^*}}$ does not vanish but is equal to $\sum_{i=1}^{d} a_i v_i u$; of course these terms do not affect the final results.

Theorem 5.26 *Consider equation (5.60) under the standing assumption that the vector fields b^j are tangent to Γ. In equation (5.60) the operators A^* and $(B^j)^*$ are those discussed in Remark 1 of section 5.5. Then:*

i) equation (5.60) is well posed and coercive in $H = L^2(D)$, $V_1^* = H^1(D)$, V_{-1}^* and V_{-2}^*; hence, it is also well posed in V_α^* for $-2 \le \alpha \le 1$;

ii) if

$$b_i^j = \frac{\partial b_i^j}{\partial v_A} = \frac{\partial c^j}{\partial v_A} = 0 \quad \text{on} \ \Gamma,$$

for all i, j, then equation (5.60) is well posed and coercive in V^*_{-3} and V^*_{-4}; hence it is also well posed in V^*_α, for $-4 \leq \alpha \leq 1$;

iii) if for all i, j, k

$$b^j_i = \frac{\partial b^j_i}{\partial x_k} = c^j = \frac{\partial c^j}{\partial x_k} = 0 \quad \text{on} \quad \Gamma,$$

and

$$\text{either} \quad \text{div } b^j = 0 \quad \text{in} \quad D, \quad \text{or} \quad \frac{\partial^2 b^j_i}{\partial x_k \partial x_l} = 0 \quad \text{on} \quad \Gamma,$$

then equation (5.60) is well posed and coercive in V^*_2 and V^*_2; hence it is also well posed in V^*_α, for $-4 \leq \alpha \leq 3$;

iv) if

$$D^\alpha b^j_i = D^\alpha c^j = 0 \quad \text{on} \quad \Gamma, \tag{5.61}$$

for all multi-indexes α such that $|\alpha| \leq 2n - 1$, then equation (5.60) is well posed in V^*_α, for $-(2n + 2) \leq \alpha \leq 2n - 1$, and it is coercive in all such spaces with integer α;

v) if, in addition to (5.61), we have

$$\text{either} \quad \text{div } b^j = 0 \quad \text{in} \quad D, \quad \text{or} \quad D^\alpha b^j_i = 0 \quad \text{on} \quad \Gamma$$

also for $|\alpha| = 2n$, then equation (5.60) is well posed in V^*_α, for $-(2n + 2) \leq \alpha \leq 2n + 1$, and it is coercive in all such spaces with integer α;

vi) finally, if the coefficients b^k_i and c^k have compact support in D, then equation (5.60) is well posed in every V^*_α, $\alpha \in \mathbf{R}$, and coercive for $\alpha \in \mathbf{Z}$.

6 Non-Homogeneous Boundary Value Problems

6.1 General Framework, Standing Assumptions, and Applications

We shall first deal with the following abstract non-homogeneous boundary value problem:

$$\begin{cases} u(t) = u_0 + \int\limits_0^t (\mathcal{A}u(s) + f^0(s)) \, ds + \sum_{j=1}^n \int\limits_0^t (B^j u(s) + f^j(s)) \, dw^j(s) \\ \gamma u(t) - g(t), \end{cases} \tag{6.1}$$

where the second equation represent some kind of boundary condition (note: in the literature on boundary value problems the symbol B is usually employed for the boundary operator, here denoted by γ; we hope this disagreement with the classical notations will not be a source of confusion). We assume to have three scales of Hilbert spaces, related one to another by $\mathcal{A}$ and γ. Let $(H) = \{H_\alpha; \alpha \in R\}$ be a Hilbert scale with norm $|.|_\alpha$ and inner product $< ., . >_\alpha$. Assume

$$\mathcal{A} : H_{\alpha+2} \rightarrow H_\alpha$$

$$B^j : H_{\alpha+1} \rightarrow H_\alpha$$

are bounded operators for all $\alpha \in \mathbf{R}$. Let $(\Gamma) = \{\Gamma_\alpha; \alpha \in \mathbf{R}\}$ another Hilbert scale, with norms $|.|_{\Gamma,\alpha}$ and inner product $< ., . >_{\Gamma,\alpha}$. Assume

$$\gamma : H_{\alpha+\rho} \rightarrow \Gamma_\alpha \tag{6.2}$$

is a bounded operator for all $\alpha > 0$. Here $\rho \in [0, 2)$ is fixed.

The definition of the third Hilbert scale, $(V) = \{V_\alpha; \alpha \in \mathbf{R}\}$, is more involved, but the final result is that of section 5.1. We set $V_2 = \{u \in H_2 : \gamma u = 0\}$, and assume that the operator $A : D(A) \subset H_0 \rightarrow H_0$ defined as

$$\begin{cases} D(A) = V_2 \\ A = \mathcal{A} \end{cases} \tag{6.3}$$

is the infinitesimal generator of an analytic semigroup of contraction in H_0, of negative type. Then we set

$$V_\alpha = D((-A)^{\frac{\alpha}{2}}),$$

with a topology equivalent to the graph topology. We finally assume that

$$V_\alpha = H_\alpha$$

for $0 \leq \alpha < \rho$, and

$$V_\alpha \subset H_\alpha,$$

V_α is a closed subspace of H_α, and that the restriction of the topology of H_α to V_α is equivalent to the topology of V_α. The properties of A in the spaces V_α which follow from this construction are those listed in section 5.1.

We now assume that the abstract *elliptic problem*

$$\begin{cases} \mathcal{A}u = 0 \\ \gamma u = g \end{cases} \tag{6.4}$$

is solvable, in the following sense: for all $\alpha > 0$ and $g \in \Gamma_\alpha$, there exists a unique $u \in H_{\alpha+\rho}$ satisfying (6.4) and depending continuously on g in these topologies. Denoting by G the *Green mapping* of this problem, defined as $Gg = u$ (g and u related by (6.4)), we have

$$G \in L(\Gamma_\alpha, H_{\alpha+\rho})$$

for all $\alpha > 0$. We also assume that this mapping has a (unique) bounded linear extension for $\alpha \leq 0$.

We do not give a direct definition of solution to (6.1) (to avoid the problem of a direct interpretation of the boundary condition). We interpret equation (6.1) in the semigroup setting by introducing the mild equation

$$u(t) = e^{tA}u_0 + \int_0^t e^{(t-s)A} f^0(s)\, ds - A \int_0^t e^{(t-s)A} Gg(s)\, ds$$

$$+ \sum_{j=1}^n \int_0^t e^{(t-s)A}(B^j u(s) + f^j(s))\, dw^j(s), \tag{6.5}$$

where all terms are well defined, and belong to $C_{\mathcal{F}}(H_0) \cap L^2_{\mathcal{F}}(H_1)$ if we have at least $u \in C_{\mathcal{F}}(H_0) \cap L^2_{\mathcal{F}}(H_1)$, $u_0 \in L^2(\mathcal{F}_0; H_0)$, $f^0 \in L^2_{\mathcal{F}}(H_{-1})$, $f^j \in L^2_{\mathcal{F}}(H_0)$, and $g \in \mathcal{G}$, where $\mathcal{G}$ will be defined at the end of section 6.2. To see that this equation is a natural version of (6.1), we refer for instance to [11].

Example 1: Second Order Dirichlet Boundary Value Problems — Let $D \subset R^d$ be a bounded open domain with regular boundary Γ. Consider the following stochastic parabolic equation in $[0, T] \times D$:

$$\begin{cases} du(t, x) = (\mathcal{A}u(t, x) + f^0(t, x))\, dt + \sum_{j=1}^n (B^j u(t, x) + f^j(t, x))\, dw^j(t), \\ u(t, x) = g(t, x), \quad t \in [0, T], x \in \Gamma, \\ u(0, x) = u_0(x), \quad x \in D, \end{cases} \tag{6.6}$$

corresponding to (5.26), where $\mathcal{A}$ and B_j are given by (5.20), (5.21), and satisfy (5.22). In this case we take:

$$H_\alpha = H^\alpha(D), \quad \Gamma_\alpha = H^\alpha(\Gamma), \quad \gamma u = u|_\Gamma, \quad \rho = \frac{1}{2}. \tag{6.7}$$

Assumption (6.2) corresponds to a well known trace result (cf. [25]). The solvability of the elliptic problem in all Sobolev spaces is proved in [25]. Thus the results of the next subsections are applicable to equation (6.6).

Example 2: Second Order Neumann Boundary Value Problems — Consider now the equation

$$\begin{cases} du(t, x) = (\mathcal{A}u(t, x) + f^0(t, x))\, dt + \sum_{j=1}^n (B^j u(t, x) + f^j(t, x))\, dw^j(t), \\ \frac{\partial u}{\partial \nu_A}(t, x) = g(t, x), \quad t \in [0, T], x \in \Gamma, \\ u(0, x) = u_0(x), \quad x \in D, \end{cases} \tag{6.8}$$

with $\mathcal{A}$ and B_j as in the previous example. The spaces H_α and Γ_α are chosen as in (6.7), but here

$$\gamma u = \frac{\partial u}{\partial \nu_A}\big|_\Gamma, \quad \rho = \frac{3}{2}. \tag{6.9}$$

Both (6.2) and the solvability of the elliptic problem in all Sobolev spaces are proved in [25]. Thus the results of the next subsections are applicable to equation (6.8) as well.

6.2 Preliminaries on Deterministic Boundary Value Problems

Consider the *deterministic* boundary value problem

$$\begin{cases} \frac{dv}{dt}(t) = \mathcal{A}v(t) + f(t) \\ \gamma v(t) = g(t) \\ v(0) = v_0. \end{cases} \tag{6.10}$$

Here we take as f, g, and v_0 certain stochastic processes and random variables, but a part from this the equation is completely deterministic. Let us say that $(v_0, f, g) \in C_\alpha$ if there exists a unique $v \in C_{\mathcal{F}}(H_\alpha) \cap L^2_{\mathcal{F}}(H_{\alpha+1})$ satisfying (6.10). A complete characterization of the space C_α is not known, but several sufficient conditions can be given, based on classical regularity results for parabolic equations (cf. [25], [23], [24], [36], for instance). We recall some sufficient conditions, taken from these references. Other conditions, given in terms of Holder-continuous processes, can be found in [11].

Example 1. Dirichlet Boundary Conditions — Assume that v is a sufficiently regular function satisfying (6.10). Then (independently of the kind of boundary condition) the following *compatibility relations* hold:

$$g(t = 0) = \gamma v_0 \tag{6.11}$$

$$\frac{\partial g}{\partial t}(t = 0) = \gamma \mathcal{A}v_0 + \gamma f(t = 0) \tag{6.12}$$

and in general

$$\frac{\partial^n g}{\partial t^n}(t = 0) = \gamma \mathcal{A}^n v_0$$

$$+ \gamma \left[\mathcal{A}^{n-1} f(t = 0) + \mathcal{A}^{n-2}\frac{\partial f}{\partial t}(t = 0) + \cdots + \frac{\partial^{n-1} g}{\partial t^{n-1}}(t = 0) \right]. \tag{6.13}$$

Denote by $H^{r;s}_{\mathcal{F}}(D), r \in \mathbf{R}, s \in \mathbf{N}$, the space of processes $L^2_{\mathcal{F}}(H^r(D)) \cap H^s_{\mathcal{F}}(0, T; L^2(D))$, where $H^s_{\mathcal{F}}(0, T; L^2(D))$ is the space of all processes f which are s-times differentiable with values in $L^2(D)$, and

$$\frac{\partial^k f}{\partial t^k} \in L^2_{\mathcal{F}}(L^2(D)), \quad k = 0, \dots, s.$$

The spaces $H^{r,s}_{\mathcal{F}}(\Gamma)$ are defined analogously. Then, with the aid of the classical results of the references listed above, one can prove the following results (which are not optimal, to keep the exposition as simple as possible).

Lemma 6.1 *Consider equation (6.10) with the notations of example 1 of section 6.1 (Dirichlet case). Let $\rho = \frac{1}{2}$, and let $[\xi]^+$ be the smallest integer $\geq \xi$, for all $\xi \in \mathbf{R}$. Then:*

i) if $f \in L^2_{\mathcal{F}}(L^2(D))$, $g \in H^{1-\rho,[\frac{1-\rho}{2}]^+}_{\mathcal{F}}(\Gamma)$, $v_0 \in L^2(\mathcal{F}_0; L^2(D))$, then there exists a unique solution $v \in C_{\mathcal{F}}(L^2(D)) \cap L^2_{\mathcal{F}}(H^1(D))$;

ii) if $f \in L^2_{\mathcal{F}}(L^2(D))$, $g \in H^{2-\rho,[\frac{2-\rho}{2}]^+}_{\mathcal{F}}(\Gamma)$, $v_0 \in L^2(\mathcal{F}_0; H^1(D))$, and the compatibility relation (6.11) is satisfied, then $v \in C_{\mathcal{F}}(H^1(D)) \cap L^2_{\mathcal{F}}(H^2(D))$;

iii) if $f \in H^{1,[\frac{1}{2}]^+}_{\mathcal{F}}(D)$, $g \in H^{3-\rho,[\frac{3-\rho}{2}]^+}_{\mathcal{F}}(\Gamma)$, $v_0 \in L^2(\mathcal{F}_0; H^2(D))$, and the compatibility relation (6.11) is satisfied, then $v \in C_{\mathcal{F}}(H^2(D)) \cap L^2_{\mathcal{F}}(H^3(D))$;

iv) in general, for $n \geq 1$, if $f \in H^{2n,[\frac{2n}{2}]^+}_{\mathcal{F}}(D)$, $g \in H^{2n+2-\rho,[\frac{2n+2-\rho}{2}]^+}_{\mathcal{F}}(\Gamma)$, $v_0 \in L^2(\mathcal{F}_0; H^{2n+1}(D))$, and the compatibility relations (6.11)–(6.13) are satisfied, then $v \in C_{\mathcal{F}}(H^{2n+1}(D)) \cap L^2_{\mathcal{F}}(H^{2n+2}(D))$;

v) if $f \in H^{2n+1,[\frac{2n+1}{2}]^+}_{\mathcal{F}}(D)$, $g \in H^{2n+3-\rho,[\frac{2n+3-\rho}{2}]^+}_{\mathcal{F}}(\Gamma)$, $v_0 \in L^2(\mathcal{F}_0; H^{2n+2}(D))$, and the compatibility relations (6.11)–(6.13) are satisfied, then $v \in C_{\mathcal{F}}(H^{2n+2}(D)) \cap L^2_{\mathcal{F}}(H^{2n+3}(D))$.

Example 2. Neumann Boundary Conditions — Similarly to the previous example, we have:

Lemma 6.2 *Consider equation (6.10) with the notations of example 2 of section 6.1 (Neumann case). Let $\rho = \frac{3}{2}$, and let $[\xi]^+$ be the smallest integer $\geq \xi$, for all $\xi \in \mathbf{R}$. Then:*

i) if $f \in L^2_{\mathcal{F}}(L^2(D))$, $g \in L^2_{\mathcal{F}}(L^2(\Gamma))$, $v_0 \in L^2(\mathcal{F}_0; L^2(D))$, then there exists a unique solution $v \in C_{\mathcal{F}}(L^2(D)) \cap L^2_{\mathcal{F}}(H^1(D))$;

ii) if $f \in L^2_{\mathcal{F}}(L^2(D))$, $g \in H^{2-\rho,[\frac{2-\rho}{2}]^+}_{\mathcal{F}}(\Gamma)$, $v_0 \in L^2(\mathcal{F}_0; H^1(D))$, then $v \in C_{\mathcal{F}}(H^1(D)) \cap L^2_{\mathcal{F}}(H^2(D))$;

iii) if $f \in H^{1,[\frac{1}{2}]^+}_{\mathcal{F}}(D)$, $g \in H^{3-\rho,[\frac{3-\rho}{2}]^+}_{\mathcal{F}}(\Gamma)$, $v_0 \in L^2(\mathcal{F}_0; H^2(D))$, and the compatibility relation (6.11) is satisfied, then $v \in C_{\mathcal{F}}(H^2(D)) \cap L^2_{\mathcal{F}}(H^3(D))$;

iv) in general, for $n \geq 1$, if $f \in H^{2n,[\frac{2n}{2}]^+}_{\mathcal{F}}(D)$, $g \in H^{2n+2-\rho,[\frac{2n+2-\rho}{2}]^+}_{\mathcal{F}}(\Gamma)$, $v_0 \in L^2(\mathcal{F}_0; H^{2n+1}(D))$, and the compatibility relations (6.11)-(6.13) are satisfied up to $n - 1$, then $v \in C_{\mathcal{F}}(H^{2n+1}(D)) \cap L^2_{\mathcal{F}}(H^{2n+2}(D))$;

v) if $f \in H_{\mathcal{F}}^{2n+1,[\frac{2n+1}{2}]^+}(D)$, $g \in H_{\mathcal{F}}^{2n+3-\rho,[\frac{2n+3-\rho}{2}]^+}(\Gamma)$, $v_0 \in L^2(\mathcal{F}_0; H^{2n+2}(D))$, and the compatibility relations (6.11)–(6.13) are satisfied (up to n), then $v \in C_{\mathcal{F}}(H^{2n+2}(D)) \cap L_{\mathcal{F}}^2(H^{2n+3}(D))$.

We use also the following notation. Let $u_0 = 0$ and $f = 0$ in equation (6.10). We denote by $\mathcal{G}$ the space of all processes g such that there exists a unique solution $v \in C_{\mathcal{F}}(H_0) \cap L_{\mathcal{F}}^2(H_1)$. Of course $\mathcal{G}$ is related to C_0.

6.3 Orientation

The basic idea is to use (6.5) in the form

$$u(t) = \phi(t) + \sum_{j=1}^{n} \int_0^t e^{(t-s)A} B^j u(s) \, dw^j(s), \qquad (6.14)$$

where

$$\phi(t) = e^{tA} u_0 + \int_0^t e^{(t-s)A} f^0(s) \, ds - A \int_0^t e^{(t-s)A} G g(s) \, ds$$

$$+ \sum_{j=1}^{n} \int_0^t e^{(t-s)A} f^j(s) \, dw^j(s). \qquad (6.15)$$

We want to apply Corollary 3.6 with $\mathcal{X} = V_\alpha$, $D(A) = V_{\alpha+2}$, $\mathcal{X}_0 = H_\alpha$, $\mathcal{Y}_0 = H_{\alpha+1}$, $\mathcal{Y} = D((-A)^{\frac{1}{2}}) = V_{\alpha+1}$. On B^j we shall impose the usual hypothesis (6.16) below. On $(u_0, f^0, \ldots, f^n, g)$ we shall require that the corresponding process ϕ defined above be in $C_{\mathcal{F}}(H_\alpha) \cap L_{\mathcal{F}}^2(H_{\alpha+1})$. Following this procedure, we divide the analysis in two cases, one almost immediate, treated in section 6.4, and the second in the direction of *optimal regularity results*. However, we note that following [11] there is a third more precise way of studing (6.5), in which this equation is not splitted at the beginning in the form (6.14)–(6.15); since the final results obtained by the latter method (although slightly stronger) are of difficult interpretation, we omit them.

6.4 Elementary Regularity

Let us set

$$\phi(t) = v(t) + m(t)$$

where

$$v(t) = e^{tA}u_0 + \int_0^t e^{(t-s)A} f^0(s)\, ds - A \int_0^t e^{(t-s)A} Gg(s)\, ds,$$

$$m(t) = \sum_{j=1}^n \int_0^t e^{(t-s)A} f^j(s)\, dw^j(s).$$

We analyze the regularity of v and m separately. This method is elementary but also restrictive, since it imposes unnecessary compatibility conditions on the data. Note that, by the above splitting, the regularity of v can be studied simply by a direct application of deterministic results (section 6.2).

Theorem 6.3 *Assume that*

$$B^j : H_{\alpha+1} \to V_\alpha, \tag{6.16}$$

$$\frac{1}{2} \sum_{j=1}^n |B^j u|_\alpha^2 \le -\eta < Au, u >_\alpha + \lambda |u|_\alpha^2, \quad u \in V_{\alpha+2}. \tag{6.17}$$

Let $(u_0, f^0, g) \in C_\alpha$, and $f^j \in L^2_{\mathcal{F}}(V_\alpha)$. Then there exists a unique solution $u \in C_{\mathcal{F}}(H_\alpha) \cap L^2_{\mathcal{F}}(H_{\alpha+1})$.

The assumption $f^j \in L^2_{\mathcal{F}}(V_\alpha)$ is imposed, instead of $f^j \in L^2_{\mathcal{F}}(H_\alpha)$ (which apparently could be sufficient to apply Corollary 3.6), because e^{tA} acts in V_α and not in H_α, in general.

Assumption (6.16) imposes the geometrical constraints on B^j discussed in section 5.

Both the assumptions on (u_0, f^0, g) and f^j are unnecessary for some reason. For instance, the assumption $f^j \in L^2_{\mathcal{F}}(V_\alpha)$ is restrictive, since it requires too many vanishing conditions of f^j on the boundary; as we shall see in the next section, a suitable form of g may compensate non-vanishing f^j.

6.5 More Refined Abstract Regularity Results

We transform ϕ in a more carefull way, in order to relax the assumptions of Theorem 6.3. The transformation of ϕ, as well as the final result, depend here more heavily on the value of α. To avoid an unreadable general statement, we treat separately three special cases, namely $0 \le \alpha < \rho$, $\rho < \alpha < 2 + \rho$, and $2 + \rho < \alpha < 4 + \rho$. The general case will then be obvious and left to the reader.

The case $0 \le \alpha < \rho$ is easy: here the result of Theorem 6.3 is already optimal (as far as the present method is concerned), since the assumption $f^j \in L^2_{\mathcal{F}}(V_\alpha)$ does not imposes any vanishing boundary condition on f^j ($V_\alpha = H_\alpha$ for such values of α, by assumption).

For the case $\rho < \alpha < 2 + \rho$ we have:

Theorem 6.4 *Let $\rho < \alpha < 2 + \rho$, and let B^j as in Theorem 6.3. Assume that g has the form*

$$g(t) = g_1(t) + M_1(t),$$

with

$$M_1(t) = \sum_{j=1}^n \int_0^t g_1^j(s) dw^j(s).$$

Assume $(u_0, f^0, g_1) \in C_\alpha$, $M_1 \in C_{\mathcal{F}}(\Gamma_{\alpha-\rho}) \cap L^2_{\mathcal{F}}(\Gamma_{\alpha+1-\rho})$, $g_1^j \in L^2_{\mathcal{F}}(\Gamma_{\alpha-\rho})$, $f^j \in L^2_{\mathcal{F}}(H_\alpha)$. Moreover, assume that g_1^j and f^j satisfy the compatibility relation (typical of the stochastic case)

$$g_1^j = \gamma f^j. \tag{6.18}$$

Then there exists a unique solution $u \in C_{\mathcal{F}}(H_\alpha) \cap L^2_{\mathcal{F}}(H_{\alpha+1})$.

Proof — Split ϕ in the form

$$\phi(t) = v_1(t) + m_1(t)$$

where

$$v_1(t) = e^{tA}u_0 + \int_0^t e^{(t-s)A} f^0(s)\, ds - A \int_0^t e^{(t-s)A} G g_1(s)\, ds,$$

$$m_1(t) = \sum_{j=1}^n \int_0^t e^{(t-s)A} f^j(s)\, dw^j(s) - A \int_0^t e^{(t-s)A} G M_1(s)\, ds$$

$$= G M_1^{\cdot}(t) + \sum_{j=1}^n \int_0^t e^{(t-s)A} [f^j(s) - G g_1^j(s)]\, dw^j(s).$$

The term v_1 is in $C_{\mathcal{F}}(H_\alpha) \cap L^2_{\mathcal{F}}(H_{\alpha+1})$ by section 6.2. Moreover, by asumption, $f^j - G g_1^j \in L^2_{\mathcal{F}}(V_\alpha)$, so that also $m_1 \in C_{\mathcal{F}}(H_\alpha) \cap L^2_{\mathcal{F}}(H_{\alpha+1})$. This completes the proof.

Finally, we study the case $2 + \rho < \alpha < 4 + \rho$.

Theorem 6.5 *Let* $2 + \rho < \alpha < 4 + \rho$, *and let* B^j *as in Theorem 6.3. Assume that g has the form*

$$g(t) = g_1(t) + M_1(t) + \int_0^t M_2(s)\, ds,$$

with

$$M_1(t) = \sum_{j=1}^n \int_0^t g_1^j(s) dw^j(s),$$

and

$$M_2(t) = \sum_{j=1}^n \int_0^t g_2^j(s) dw^j(s).$$

Assume $(u_0, f^0, g_1) \in C_\alpha$, $M_1 \in C_{\mathcal{F}}(H_{\alpha-\rho}) \cap L^2_{\mathcal{F}}(H_{\alpha+1-\rho})$, $M_2 \in C_{\mathcal{F}}(H_{\alpha-2-\rho}) \cap L^2_{\mathcal{F}}(H_{\alpha+1-2-\rho})$, $g_1^j \in L^2_{\mathcal{F}}(\Gamma_{\alpha-\rho})$, $g_2^j \in L^2_{\mathcal{F}}(\Gamma_{\alpha-2-\rho})$, $f^j \in L^2_{\mathcal{F}}(H_\alpha)$. *Moreover, assume that* g_1^j, g_2^j, *and* f^j *satisfy the compatibility relation (typical of the stochastic case)*

$$g_1^j = \gamma f^j, \tag{6.19}$$

$$g_2^j = \gamma A f^j. \tag{6.20}$$

Then there exists a unique solution $u \in C_{\mathcal{F}}(H_\alpha) \cap L^2_{\mathcal{F}}(H_{\alpha+1})$.

Proof — Split ϕ in the form

$$\phi(t) = v_1(t) + m_2(t)$$

where v_1 is the same of the previous proof, and

$$m_2(t) = GM_1(t) + \sum_{j=1}^n \int_0^t e^{(t-s)A}[f^j(s) - Gg_1^j(s)]\, dw^j(s)$$

$$-A \int_0^t e^{(t-s)A} GM_2(s)\, ds$$

$$= GM_1(t) + A^{-1}GM_2(t) + \sum_{j=1}^n \int_0^t e^{(t-s)A}[f^j(s) - Gg_1^j(s) - A^{-1}Gg_2^j(s)]\, dw^j(s).$$

The proof is now analogous to the previous one, and is left to the reader.

7   Existence and Regularity of Stochastic Flows

7.1   First Abstract Theorems of Existence and Regularity

In this section we use notations and results of section 5.1. Thus, we consider equation (5.2) of that section. Moreover, we refer to section 2 for the concepts of random field, regular version, stochastic flow, and the first results concerning them.

To express in a more natural way certain evolution properties of the flows, let us consider (just in this subsection) an equation of the form (5.2) starting at a generic time $s \geq 0$:

$$\begin{cases} du(t) = Au(t)\, dt + \sum_{j=1}^{\infty} B^j u(t)\, dw^j(t), & t \in [s, T], \\ u(s) = x, \end{cases} \qquad (7.1)$$

interpreted in the mild form

$$u(t) = e^{(t-s)A} x + \sum_{j=1}^{\infty} \int_s^t e^{(t-\sigma)A} B^j u(\sigma)\, dw^j(\sigma). \qquad (7.2)$$

If U and Z are Hilbert spaces, we denote by $L_2(U, Z)$ the space of *Hilbert-Schmidt operators* from U to Z.

Definition 7.1: *In this section we say that equation (7.1) generates a stochastic flow in a space V_α if it is well posed in V_α (cf. Definition 5.1) and for all $s < t \in [0, T]$ the random field $x \mapsto u(t, s, x)$ (solution of (7.1) at time t), from V_α to $L^2(\mathcal{F}_t; V_\alpha)$, has a Hilbert-Schmidt modification $\{\phi_{s,t}(\omega); \omega \in \Omega\} \subset L_2(V_\alpha)$ (cf. section 2). Sometimes, we may emphasize the Hilbert-Schmidt property saying that equation (7.1) generates a Hilbert-Schmidt stochastic flow in V_α. If $s = 0$, $\phi_{s,t}(\omega)$ will be simply denoted by $\phi_t(\omega)$.*

Note that (as suggested us by the referee) the operator $(-A)^{-\delta/2}$ is Hilbert-Schmidt in V_α, for some α and $\delta > 0$, if and only if it is Hilbert-Schmidt in H. Indeed, $(-A)^\alpha$ maps isometrically V_α onto H. So, if $\{e_n\}$ is a complete orthonormal system in H, then $\{(-A)^{-\alpha} e_n\}$ is a complete orthonormal system in V_α. Let $(-A)^{-\delta/2}$ be Hilbert-Schmidt in H. Then

$$\sum_{n=1}^{\infty} \|(-A)^{-\delta/2}(-A)^{-\alpha} e_n\|_{V_\alpha}^2$$

$$= \sum_{n=1}^{\infty} \|(-A)^\alpha (-A)^{-\delta/2}(-A)^{-\alpha} e_n\|_H^2$$

$$= \sum_{n=1}^{\infty} \|(-A)^{-\delta/2}(-A)^\alpha (-A)^{-\alpha} e_n\|_H^2$$

$$= \sum_{n=1}^{\infty} \|(-A)^{-\delta/2} e_n\|_H^2 < \infty.$$

The converse implication is analogous. We shall use this fact in the sequel.

Theorem 7.2 *Let $s \in [0, T]$ be given. Assume that for a given $\alpha \in \mathbf{R}$ equation (7.1) is well posed in both V_α and $V_{\alpha-\delta}$, where $\delta > 0$ is such that $(-A)^{-\delta/2}$ is a Hilbert-Schmidt operator in H. Then equation (7.1) generates a Hilbert-Schmidt stochastic flow $\phi_{s,t}(\omega)$ in V_α.*

Proof — The operator $(-A)^{\delta/2}$ is a bounded linear operator from V_α to $V_{\alpha-\delta}$. Hence, recalling the second part of Theorem 5.4, for all given $s < t \in [0, T]$, the mapping $x \mapsto u(t, s, (-A)^{\delta/2}x)$ is bounded from V_α to $L^2(\mathcal{F}_t; V_\alpha)$. Therefore, the mapping $x \mapsto u(t, s, x) = u(t, s, (-A)^{\delta/2} (-A)^{-\delta/2}x)$ is Hilbert-Schmidt from V_α to $L^2(\mathcal{F}_t; V_\alpha)$. To complete the proof, it is sufficient to apply Lemma 2.3.

A slight extension of this Theorem provides a regularity result for the flow:

Theorem 7.3 *Let $s \in [0, T]$ and $\alpha \leq \beta$ be given. Assume that equation (7.1) is well posed in both V_β and $V_{\alpha-\delta}$, where $\delta > 0$ is such that $(-A)^{-\delta/2}$ is a Hilbert-Schmidt operator in H. Then the stochastic flow $\phi_{s,t}(\omega)$ generated by equation (7.1) in V_α has the regularity property $\phi_{s,t}(\omega) \in L_2(V_\alpha, V_\beta)$ P-a.s., for all $s < t \in [0, T]$.*

We omit the proof that is completely analogous to the previous one.

Under the assumptions of Theorem 7.2, and at least in the case of a finite number m of Wiener processes (i.e. $B^j = 0$ for $j > m$), it is possible to prove some properties of uniformity in t of $\phi_{s,t}(\omega)$. For instance, fixed $s \in [0, T)$,

$$t \mapsto \phi_{s,t}(\omega) \text{ is continuous from } (s, T] \text{ to } L_2(V_\alpha), \quad P - a.s.$$

$$\forall x \in V_\alpha, \quad \phi_{s,.}(\omega)x = u(., s, x)(\omega) \quad \text{on} \quad [s, T], \quad P - a.s.$$

$$E \sup_{\varepsilon \leq t \leq T} |\phi_{s,t}(\omega)|^2_{L_2(V_\alpha)} < \infty, \quad \forall \varepsilon \in (s, T).$$

Moreover, one can prove that $\phi_{s,t}(\omega)$ satisfies the *evolution property*

$$\phi_{s,t}(\omega) = \phi_{r,t}(\omega)\phi_{s,r}(\omega), \quad s \leq r \leq t, \quad P - a.s.$$

uniformly in $t \in [r, T]$, for all given $s \leq r$. Similarly, if $T = +\infty$, let Ω be the canonical Wiener space $\Omega = \{f \in C([0, \infty); R^m), f(0) = 0\}$, and Θ_t be the canonical shift on Ω defined as $(\Theta_t\omega)(s) = \omega(t + s)$ $(t, s \geq 0)$; then, denoting $\phi_{0,t}$ simply by ϕ_t, the following *cocycle property* can be proved:

$$\phi_t(\Theta_s\omega)\phi_s(\omega) = \phi_{t+s}(\omega), \quad P - a.s.$$

uniformly in $t \geq 0$, for all fixed $s \geq 0$. The proof of these properties can be found in [4] and [12]. They are important, for instance, in the applications of the notion of stochastic flow to the analysis of *Lyapunov exponents* and asymptotic behaviour for equation (7.1) (see [12]).

7.2 A Pathwise Green Formula

Always with the notations of section 5.1, we consider now simultaneously the two adjoint (or dual) equations

$$du = Au\,dt + \sum_{j=1}^{\infty} B^j u\,dw^j, \quad u(0) = u_0, \tag{7.3}$$

and

$$dz = A^*z\,dt + \sum_{j=1}^{\infty} B^{j*}z\,d\tilde{w}^j, \quad z(0) = z_0, \tag{7.4}$$

both on $[0, T]$, with $\tilde{w}^j(t) = w(T) - w(T-t)$. Here the adjoint operators are defined with respect to the inner product in H.

We assume that (7.3) and (7.4) are well posed and coercive in H, i.e. that

$$\begin{cases} B^j \in L(D((-A)^{\frac{1}{2}}), H), \\ \frac{1}{2}\sum_{j=1}^{\infty} |B^j u|_H^2 \leq -\eta <Au, u>_H +\lambda|u|_H^2, \quad u \in D(A), \\ \sum_{j=1}^{\infty} |B^j u|_H^2 \leq c|u|^2_{D((-A)^{\frac{1}{2}})}, \quad u \in D((-A)^{\frac{1}{2}}), \end{cases} \tag{7.6}$$

and

$$\begin{cases} B^{j*} \in L(D((-A^*)^{\frac{1}{2}}), H), \\ \frac{1}{2}\sum_{j=1}^{\infty} |B^{j*}|_H^2 \leq -\eta <A^*u, u>_H +\lambda|u|_H^2, \quad u \in D(A^*), \\ \sum_{j=1}^{\infty} |B^{j*}u|_H^2 \leq c|u|^2_{D((-A^*)^{\frac{1}{2}})}, \quad u \in D((-A^*)^{\frac{1}{2}}), \end{cases} \tag{7.6}$$

for some constants $\eta \in (0, 1)$, $\lambda \geq 0$, and $c > 0$.

The following result holds in a generic Hilbert space, so that, for instance, one could take a certain V_α in place of H, with the understanding that the adjoints are in the sense of V_α.

Particular forms of the following Lemma are known in the literature (cf. for instance [28]), but we give the proof for completeness since the present generality is not covered by the known results.

Lemma 7.4 *Under the assumptions (7.5)–(7.6), for every u_0 and z_0 in H,*

$$<u(T), z_0>_H=<u_0, z(T)>_H \quad P-a.s. \tag{7.7}$$

A formal proof proceeds as follows: rewriting (7.3) and (7.4) in Stratonovich form

$$du = (A - \frac{1}{2}\sum_{j=1}^{\infty} B^j)u\,dt + \sum_{j=1}^{\infty} B^j u \circ dw^j, \quad u(0) = u_0, \tag{7.8}$$

and

$$dz = (A^* - \frac{1}{2}\sum_{j=1}^{\infty} B^{j*})z\,dt + \sum_{j=1}^{\infty} B^{j*}z\,o\,d\tilde{w}^j, \quad z(0) = z_0, \tag{7.9}$$

we may formally apply classical differential calculus and obtain (as it is easily verified)

$$\frac{d}{dt} < u(t), z(T - t) >_H = 0. \tag{7.10}$$

Let us now give a rigorous proof by Euler approximation method.

Proof. Step 1 — It is sufficient to prove the Lemma in the case when the operators B^j are bounded in H. Indeed, introducing the Yosida approximations $J_n = n(n - A)^{-1}$ (cf. [29]), we can consider the approximating equations

$$du_n = Au_n\,dt + \sum_{j=1}^{\infty} B_n^j u_n\,dw^j, \quad u_n(0) = u_0, \tag{7.11}$$

and

$$dz_n = A^* z_n\,dt + \sum_{j=1}^{\infty} B_n^{j*} z_n\,d\tilde{w}^j, \quad z_n(0) = z_0, \tag{7.12}$$

where $B_n^j = B^j J_n$. By Lemma 3.4, u_n and z_n converge to u and z in $C_{\mathcal{F}}(H)$. Hence, if (7.7) is proved for u_n and z_n, it follows in the limit for u and z.

Step 2 — It is now sufficient to prove the Lemma when also the operator A is bounded. Assuming that the operators B^j in equations (7.3) and (7.4) are bounded (by step 1), consider the approximating equations

$$du_n = A_n u_n\,dt + \sum_{j=1}^{\infty} B^j u_n\,dw^j, \quad u_n(0) = u_0, \tag{7.13}$$

and

$$dz_n = A_n^* z_n\,dt + \sum_{j=1}^{\infty} B^{j*} z_n\,d\tilde{w}^j, \quad z_n(0) = z_0, \tag{7.14}$$

where $A_n = A J_n$. It is well known that equations (7.13) and (7.14) have unique solutions in $C_{\mathcal{F}}(H)$, and that u_n and z_n converge to u and z in $C_{\mathcal{F}}(H)$ (cf. [17] and [9]). Hence we conclude as above.

Step 3 — Assume now that A and B^j are bounded. For any fixed n, let $h = \frac{T}{n}$, $t_k = kh$ for $k = 0, \ldots, n$, and let $u_n(t)$, $z_n(t)$ be functions defined for $t = t_k$ as follows: $u_n(t_0) = u_0$, $z_n(t_0) = z_0$, and

$$u_n(t_{k+1}) = u_n(t_k) + hAu_n(t_k) + \sum_{j=1}^{\infty} B^j u_n(t_k)(w(t_{k+1}) - w(t_k)), \qquad (7.15)$$

$$z_n(t_{k+1}) = z_n(t_k) + hA^* z_n(t_k) + \sum_{j=1}^{\infty} B^{j*} z_n(t_k)(\tilde{w}(t_{k+1}) - \tilde{w}(t_k)) \qquad (7.16)$$

for $k = 1, \ldots, n$. A strightforward computation gives

$$< u_n(t_n), z_n(t_0) >_H = < u_n(t_{n-1}), z_n(t_1) >_H = \cdots = < u_n(t_0), z_n(t_n) >_H . \qquad (7.17)$$

(using the identity $\tilde{w}(t_{k+1}) - \tilde{w}(t_k) = w(T - t_k) - w(T - t_{k+1}) = w(t_{n-k}) - w(t_{n-k-1})$). Hence, in particular,

$$< u_n(T), z_0 >_H = < u_0, z_n(T) >_H, \qquad P - a.s. \qquad (7.18)$$

Step 4 — Finally, let u and z be the solutions of equations (7.3) and (7.4) with bounded operators A and B^j. We shall prove that the Euler approximations u_n and z_n of step 3, suitably extended out of the grid where thy are originally defined, converge to u and z in $C_{\mathcal{F}}(H)$. Thus (7.7) follows from (7.18). Of course, it is sufficient to give the proof of the convergence of u_n to u, since the equations for z_n and z are of the same form.

For a fixed $n \in \mathbb{N}$ and all $s \in [0, T]$, set $[s]_n = \max\{t_k; t_k \leq s\}$, where $t_k = k\frac{T}{n}$ as above. Moreover, define

$$u_n(t) = u_0 + \int_0^t Au_n([s]_n) \, ds + \sum_{j=1}^{\infty} \int_0^t B^j u_n([s]_n) \, dw^j(s).$$

One checks immediately that this process extends the process u_n previously defined only for $t = t_0, \ldots, t_n$. Moreover, note that

$$u(t) = u_0 + \int_0^t Au([s]_n) \, ds + \sum_{j=1}^{\infty} \int_0^t B^j u([s]_n) \, dw^j(s) + R_n(t),$$

where

$$R_n(t) = \int_0^t A(u(s) - u([s]_n)) \, ds + \sum_{j=1}^{\infty} \int_0^t B^j(u(s) - u([s]_n)) \, dw^j(s).$$

Denote by c a generic constant independent of n. From the previous equations we have, for all $t \in [0, T]$,

$$E \sup_{0 \le \sigma \le t} \|u(\sigma) - u_n(\sigma)\|_H^2$$

$$\le c \int_0^t E \sup_{0 \le \sigma \le s} \|u(\sigma) - u_n(\sigma)\|_H^2 \, ds + E \sup_{0 \le \sigma \le t} \|R_n(\sigma)\|_H^2,$$

so that, if we prove that

$$E \sup_{0 \le t \le T} \|R_n(t)\|_H^2 \to 0 \qquad (7.19)$$

as $n \to \infty$, then, by Gronwall Lemma, we obtain

$$E \sup_{0 \le t \le T} \|u(t) - u_n(t)\|_H^2 \to 0$$

as $n \to \infty$, which yields the claimed convergence result.

From the definition of R_n we have

$$E \sup_{0 \le t \le T} \|R_n(t)\|_H^2 \le cE \int_0^T \|u(t) - u([t]_n)\|_H^2 dt,$$

and from the identity

$$u(t) - u([t]_n) = \int_{[t]_n}^t A u(s) \, ds + \sum_{j=1}^{\infty} \int_{[t]_n}^t B^j u(s) \, dw^j(s)$$

we have

$$E\|u(t) - u([t]_n)\|_H^2 \le ch_n \sup_{0 \le t \le T} E\|u(t)\|_H^2,$$

where $h_n = \frac{T}{n}$. Collecting these inequalities we obtain (7.19), and the proof is complete.

7.3 Transposition of the Adjoint Flow

We refer again to the notations of section 5.1. In particular, we recall that $V_\alpha^* = D((-A^*)^{\alpha/2})$.

Theorem 7.5 *Assume that both systems (7.3) and (7.4) are well posed and coercive in H (cf. (7.5)–(7.6)). Moreover, assume that for some $\alpha \geq 0$ the adjoint equation (7.4) generates a Hilbert-Schmidt stochastic flow $\phi_t^*(\omega)$ in V_α^* (cf. Definition 7.1). Let $(\phi_t^*(\omega))' \in L_2(V_{-\alpha})$ be the dual of $\phi_t^*(\omega)$. Then:*

i) if equation (7.3) is well posed in $V_{-\alpha}$, then $(\phi_t^*(\omega))'$ is the stochastic flow associated to equation (7.3) in $V_{-\alpha}$;

ii) if equation (7.3) generates a stochastic flow $\phi_t(\omega)$ in some $V_{-\beta}$ with $\beta \leq \alpha$, then P-a.s. $(\phi_t^*(\omega))'$ is the unique extension of $\phi_t(\omega)$ to $V_{-\alpha}$;

iii) in the situation of part ii), if in addition equation (7.3) is well-posed in some $V_{-\gamma}$ with $-\alpha \leq -\gamma \leq -\beta$, then it generates a stochastic flow in $V_{-\gamma}$, which is the unique extension to $V_{-\gamma}$ of $\phi_t(\omega)$.

Lemma 7.6 *Let $Y \subset X$ be two separable Hilbert spaces with compact dense embedding. Identifying X with its dual X', we have $Y \subset X \subset Y'$. Then:*

i) if T is a Hilbert-Schmidt operator in Y, then its dual operator T' is Hilbert-Schmidt in Y';

ii) if $T \in L(X) \cap L(Y)$ is a Hilbert-Schmidt operator both in X and Y, then it is Hilbert-Schmidt in $[X, Y]_\theta$ for all $\theta \in (0, 1)$ (cf. section 3.5).

Proof of the Lemma. Step 1 — Let us collect some preliminary facts. Let $\Lambda : D(\Lambda) = Y \subset X \to X$ be a selfadjoint and positive operator in X, continuous from Y to X, with bounded inverse $\Lambda^{-1} : X \to Y$ (see [25], Chapter 1). The topology in $D(\Lambda)$ defined by $||x||_Y := |\Lambda x|_X$ is equivalent to the original topology of Y. Since Λ^{-1} is compact in X, there exists a complete orthonormal system $\{e_j\}$ in X such that

$$\Lambda x = \sum_{j=1}^\infty \lambda_j < x, e_j > e_j \quad \forall x \in Y,$$

where λ_j are the eigenvalues og Λ, $\lambda_j > 0$, $\lambda_j \to +\infty$ as $j \to \infty$. An important fact is that $\{\Lambda^{-1}e_j\}$ is a complete orthonormal system in Y, with respect to $||.||_Y$.

One readily checks that Λ extends to an isomorphism from X to Y', that $||x||_{Y'} := |\Lambda^{-1}x|_X$ is an equivalent topology in Y', and that $\{\Lambda e_j\}$ is a complete orthonormal system in Y', with respect to $||.||_{Y'}$.

Step 2 — Consider part i). T' is bounded in Y', and

$$\sum_{i,j=1}^\infty << T'\Lambda e_i, \Lambda e_j >>_{Y'}^2$$

$$= \sum_{i,j=1}^\infty < \Lambda^{-1}T'\Lambda e_i, e_j >_X^2 = \sum_{i,j=1}^\infty < \Lambda\Lambda^{-1}e_i, \Lambda T\Lambda^{-1}e_j >_X^2$$

$$= \sum_{i,j=1}^{\infty} << \Lambda^{-1}e_i, T\Lambda^{-1}e_j >>_Y^2 < \infty.$$

This proves i).

Step 3 — Consider now part ii). T is a bounded operator in $[X, Y]_\theta$ for all $\theta \in (0, 1)$. Moreover, recall that there exists a constant $c_\theta > 0$ such that

$$|x|_{[X,Y]_\theta} \le c_\theta |x|_X^{1-\theta} |x|_Y^\theta, \quad \forall x \in Y.$$

Since $[X, Y]_\theta = D(\Lambda^\theta)$ with equivalent norms, $\{\lambda_j^{-\theta} e_j\}$ is a complete orthonormal system in $[X, Y]_\theta$, with respect to $||.||_\theta := |\Lambda^\theta x|_X$. Then, there exists a constant c_θ' such that

$$\sum_{i=1}^{\infty} ||T\frac{e_j}{\lambda_j^\theta}||_\theta^2 = \sum_{i=1}^{\infty} \frac{1}{\lambda_j^{2\theta}} ||Te_j||_\theta^2$$

$$\le c_\theta' \sum_{i=1}^{\infty} \frac{1}{\lambda_j^{2\theta}} |Te_j|_X^{2-2\theta} |Te_j|_Y^{2\theta} \le c_\theta' \sum_{i=1}^{\infty} |Te_j|_X^{2-2\theta} ||T\frac{e_j}{\lambda_j}||_Y^{2\theta}$$

$$\le c_\theta' \left(\sum_{i=1}^{\infty} |Te_j|_X^2 \right)^{1-\theta} \left(\sum_{i=1}^{\infty} ||T\frac{e_j}{\lambda_j}||_Y^2 \right)^\theta < \infty$$

(recall that $\{\frac{e_j}{\lambda_j}\}$ is a complete orthonormal system in Y). Hence T is Hilbert-Schmidt in $[X, Y]_\theta$. The proof of the Lemma is complete.

Proof of Theorem 7.5 — Let us introduce some preliminars. By the Lemma, the operators $(\phi_t^*(\omega))'$ are Hilbert-Schmidt in $V_{-\alpha}$. From Lemma 7.4, we know that for all $u_0 \in H$ and $z_0 \in V_\alpha^* \subset H$ (t is fixed) $< u(t; u_0), z_0 >_0 = < u_0, z(t; z_0) >_0$ P-a.s., where $u(t; u_0)$ and $z(t; z_0)$ are the solutions in H of (7.3) and (7.4) respectively. By uniqueness, $z(t; z_0)$ is also the solution in V_α^*, and from the existence of the flow in V_α^* we have $z(t; z_0) = \phi_t^*(\omega)z_0$ P-a.s.. Hence, reacalling the duality between the spaces V_α^* and $V_{-\alpha}$, for all $u_0 \in H$ and $z_0 \in V_\alpha^*$ we have

$$< u(t; u_0), z_0 >_{V_{-\alpha}, V_\alpha^*} = < (\phi_t^*(\omega))'u_0, z_0 >_{V_{-\alpha}, V_\alpha^*} \quad P - a.s.$$

Since V_α^* is separable, we easily deduce that for all $u_0 \in H$

$$u(t; u_0) = (\phi_t^*(\omega))'u_0 \quad P - a.s., \tag{7.20}$$

where we have to recall that the right-hand-side belongs to $V_{-\alpha}$, and not to H in general.

Proof of Part i) — From the continuous dependence of $u(t; u_0)$ on u_0 in $V_{-\alpha}$, the continuity of $(\phi_t^*(\omega))'$ in $V_{-\alpha}$, and the density of H in $V_{-\alpha}$, we may extend the identity (7.20) to all $u_0 \in V_{-\alpha}$, proving that $(\phi_t^*(\omega))'$ is the stochastic flow associated to equation (7.3) in this space.

Part ii) — Consider first the case $-\beta < 0$. The identity

$$u(t, u_0) = \phi_t(\omega)u_0 \quad P - a.s.$$

holds for all $u_0 \in V_{-\beta}$, hence for all $u_0 \in H$. Thus, from (7.20)

$$\phi_t(\omega)u_0 = (\phi_t^*(\omega))'u_0 \quad P - a.s., \tag{7.21}$$

for all $u_0 \in H$, where the right-hand-side belongs to $V_{-\alpha}$ and the left-hand-side to $V_{-\beta}$. By the density of H in $V_{-\beta}$, (7.21) holds true for all $u_0 \in V_{-\beta}$; hence the separability of $V_{-\beta}$ easily implies that P-a.s. $(\phi_t^*(\omega))'$ extends $\phi_t(\omega)$.

If $-\beta \geq 0$, then $u(t, u_0) = \phi_t(\omega)u_0$ P-a.s., for $u_0 \in V_{-\beta}$, so that (7.21) holds true for $u_0 \in V_{-\beta}$ directly. We conclude as above by the separability of $V_{-\beta}$.

Part iii) — The operators $\phi_t(\omega)$ are Hilbert-Schmidt both in $V_{-\alpha}$ and $V_{-\beta}$, by the previous results. Hence, from Lemma 7.6, they are Hilbert-Schmidt in every $V_{-\gamma}$, with $-\alpha \leq -\gamma \leq -\beta$. Now, $V_{-\beta}$ is dense in $V_{-\gamma}$, $u(t, u_0)$ depend continuously on u_0 in $V_{-\gamma}$, and the operators $\phi_t(\omega)$ are continuous in $V_{-\gamma}$. Therefore, the identity $u(t, u_0) = \phi_t(\omega)u_0$ which holds true P-a.s., for $u_0 \in V_{-\beta}$, can be extended by continuity to all $u_0 \in V_{-\gamma}$. This completes the proof of the Theorem.

In the applications we shall use the following results, which is a corollary of Theorems 7.2 and 7.5.

Theorem 7.7 *Assume that equations (7.3) and (7.4) are well posed and coercive in H (cf. (7.5)–(7.6)). Moreover assume that for some $\alpha > 0$ and $s \geq 0$, equation (7.3) is well posed in V_α and in $V_{-\alpha-s}$, and equation (7.4) is well posed in V_α^* and in $V_{-\alpha-s}^*$. Finally, assume that $(-A)^{-\alpha-\frac{s}{2}}$ and $(-A^*)^{-\alpha-\frac{s}{2}}$ are Hilbert-Schmidt operators in H.*

Then equation (7.3) generates a Hilbert-Schmidt stochastic flow in V_γ, for all $-\alpha \leq -\gamma \leq \alpha$ (similarly, equation (7.4) generates a Hilbert-Schmidt stochastic flow in V_γ^*, for all $-\alpha \leq -\gamma \leq \alpha$).

The next results, which also readily follows from Theorems 7.2 and 7.5, will not be used in our applications, but is conceptually of interest, as remarked below.

Theorem 7.8 *Assume that equations (7.3) and (7.4) are well posed and coercive in H (cf. (7.5)–(7.6)). Moreover assume that for some $\alpha > 0$ equation (7.3) is well posed in V_α and equation (7.4) is well posed in V_α^*. Finally, assume that $(-A)^{-\frac{\alpha}{2}}$ and $(-A^*)^{-\frac{\alpha}{2}}$ are Hilbert-Schmidt operators in H.*

Then equation (7.3) generates a Hilbert-Schmidt stochastic flow $\phi_t(\omega)$ in V_γ, for all $0 \leq -\gamma \leq \alpha$ (similarly, equation (7.4) generates a Hilbert-Schmidt stochastic flow $\phi_t^*(\omega)$ in V_γ^*, for all $0 \leq -\gamma \leq \alpha$).

Moreover, for all $-\alpha \leq \gamma \leq 0$, $\phi_t(\omega)$ has a Hilbert-Schmidt extension to V_γ, P-a.s.; thus, if in addition equation (7.3) is well posed in $V_{-\alpha}$, then it generates a stochastic flow in every V_γ, $-\alpha \leq \gamma \leq 0$.

The significance of this results is that it allows to construct the stochastic flow in H without studying explicitly the regularity in negative-order spaces, as required by Theorems 7.2 and

7.7. It is also a way to extend to the stochastic case the classical transposition method known in the regularity theory of deterministic partial differential equations (cf. [25]).

7.4 Applications

We discuss existence and regularity of stochastic flows for the boundary value problems treated in section 5. The following Theorems contain only some of the statements which follows from the regularity results of section 5 and the abstract results of this section.

7.4.1 Dirichlet Boundary Value Problem

Let us consider the second order parabolic equation (5.26) with Dirichlet boundary condition. For this equation, it has been proved in [16], by means of a Feynman-Kac representation formula, that the stochastic flow exists in $L^2(D)$ without any restriction on the coefficients b_i^j and c^j. Some regularity of the flow has been obtained in [16] under certain restriction. The approach of the present paper, in contrast, does not provide the existence in $L^2(D)$ in such generality, but yields a wide number of additional existence results in different Sobolev spaces and regularity results that have not been proved by the Feynman-Kac approach.

The proof of the following Theorem is a strightforward application of the regularity results of Theorem 5.20, of Theorem 7.3 and Theorem 7.7 with $s = 1$, and of the fact that $(-A)^{-\frac{r}{2}}$ and $(-A^*)^{-\frac{r}{2}}$ are Hilbert-Schmidt in H if $r > \frac{d}{2}$, where d is the space dimension (this corresponds to the fact that the embedding of $H^r(D)$ in $L^2(D)$ is Hilbert-Schmidt for such r).

Theorem 7.9 *Consider equation (5.26), with A and B^j defined by (5.20)–(5.21) with C^∞-coefficients in $\overline{D}$, and satisfying the joint ellipticity condition (5.22). Then:*

i) if the vector fields b^j are tangent to Γ, and the space dimension d is less than or equal to 9, then there exists the stochastic flow in V_α for all $-2 \le \alpha \le 2$; moreover, if $d \le 5$, such flow has the regularity property:

$$\phi_t(\omega) \in L_2(H, V_2) \quad P - a.s.;$$

ii) if

$$D^\alpha b_l^k = 0, \quad D^\beta c^k = 0 \quad \text{on} \quad \Gamma,$$

for all multi-indexes α and β with $|\alpha| \le 2n$ and $|\beta| \le 2n - 1$, and for all the values of k, l, and if the space dimension d is less or equal to $8n + 9$, then there exists the stochastic flow in V_α for all $-(2n + 2) \le \alpha \le 2n + 2$; moreover, if $d \le 4n + 5$, such flow has the regularity property:

$$\phi_t(\omega) \in L_2(H, V_{2n+2}) \quad P - a.s.;$$

iii) finally, if the coefficients b_i^k and c^k have compact support in D, then for all space dimension there exists the stochastic flow in every space V_α, $\alpha \in \mathbf{R}$, and for every $\beta \geq \alpha$ the flow has the regularity property:

$$\phi_t(\omega) \in L_2(V_\alpha, V_\beta) \quad P - a.s..$$

7.4.2 Equations of Order $2m$ with Dirichlet Boundary Conditions

As an example, consider problem (5.44) of section 5.4. The proof is similar to that of Theorem 7.9 (using Theorem 5.22).

Theorem 7.10 *If the coefficients of the differential operators B^j of equation (5.44) have compact support in D, then for all space dimension there exists the stochastic flow associated to (5.44) in every space V_α, $\alpha \in \mathbf{R}$. Moreover, for every $\beta \geq \alpha$ this flow has the regularity property:*

$$\phi_t(\omega) \in L_2(V_\alpha, V_\beta) \quad P - a.s..$$

7.4.3 Neumann Boundary Condition

Consider now the Neumann boundary value problem (5.46). The Feynman-Kac approach has not been developed yet in the Neumann case, and no results on flows for Neumann boundary value problems have been proved by methods different from the present one.

The proof of the following Theorem is a strightforward application of Theorem 5.24, Theorem 7.3 and Theorem 7.7 with $s = 1$, and of the same Hilbert-Schmidt properties mentioned in section 7.4.1.

Theorem 7.11 *Consider equation (5.46) with $\mathcal{A}$ and B^j defined by (5.20)–(5.21) (C^∞-coefficients in $\overline{D}$ satisfying the joint ellipticity condition (5.22)). Then:*

i) in one space dimension there exists the stochastic flow in $H^1(D)$;

ii) if the vector fields b^j are tangent to Γ, and the space dimension is less or equal to 5, then there exists the stochastic flow in V_α for $-1 \leq \alpha \leq 1$; moreover, if $d \leq 3$, this flow has the regularity property:

$$\phi_t(\omega) \in L_2(H, V_1) \quad P - a.s.;$$

iii) if

$$D^\alpha b_i^j = D^\alpha c^j = 0 \quad \text{on} \quad \Gamma,$$

for all multi-indexes α such that $|\alpha| \leq 2n - 1$, and

$$\text{either} \quad \text{div } b^j = 0 \quad \text{in} \quad D, \quad \text{or} \quad D^\alpha b_i^j = 0 \quad \text{on} \quad \Gamma \qquad (7.22)$$

for $|\alpha| = 2n$, *and if the space dimension is less or equal to* $8n + 5$, *then there exists the stochastic flow in* V_α *for* $-(2n + 1) \leq \alpha \leq 2n + 1$; *moreover, if* $d \leq 4n + 3$, *the flow has the regularity property:*

$$\phi_t(\omega) \in L_2(H, V_{2n+1}) \quad P - a.s.;$$

iv) finally, if the coefficients b_i^k *and* c^k *have compact support in* D, *then for all space dimension there exists the stochastic flow in every space* V_α, $\alpha \in \mathbf{R}$, *and for every* $\beta \geq \alpha$ *such flow has the regularity property:*

$$\phi_t(\omega) \in L_2(V_\alpha, V_\beta) \quad P - a.s..$$

8   An Alternative Approach

In this final section we show, by a different method, that the Dirichlet and Neumann boundary balue problems for second order stochastic parabolic equations generate a stochastic flow in $L^2(D)$, without some of the geometric restrictions imposed by the abstract approach of the previous sections. In contrast, this result is restricted to such particular classes and to L^2-spaces, at present, and does not give us any regularity of the flow. Anyway, it shows that an optimal theory has still to be devised.

Let D be a regular bounded open domain of $\mathbf{R}^d$ and let $a_{ij}, a_i, a, b_i^j, c^j, i = 1, \ldots, d$, $j = 1, \ldots, n$, be real valued functions in $\overline{D}$, that for sake of simplicity we assume of class $C^\infty(\overline{D})$. Let $\mathcal{A}$ be the second order strongly elliptic operator in D defined by (5.20) and let B^j be the first order differential operators in D, with $j = 1, \ldots, n$, defined by (5.21). We assume that there exists $\rho > 0$ such that the coercivity condition (5.22) is fulfilled.

Consider the Dirichlet boundary value problem (5.26) and the Neumann one (5.46).

Theorem 8.1 *For the Dirichlet problem (5.26) there exists a stochastic flow in* $L^2(D)$.

Theorem 8.2 *For the Neumann problem (5.46), assume that all the vector fields* $b^j(x)$ *are tangent to the boundary. Then there exists a stochastic flow in* $L^2(D)$.

Proof of Theorems 1.1–1.2.   Let $\tilde{B}^j$ and $\hat{B}^j$ be the first order differential operators associated to B^j, defined as

$$\tilde{B}^j u(x) = \sum_{i=1}^d b_i^j(x) \frac{\partial u(x)}{\partial x_i} + \tilde{c}^j(x) u(x)$$

$$\hat{B}^j u(x) = \sum_{i=1}^d b_i^j(x) \frac{\partial u(x)}{\partial x_i} + (c^j(x) + \tilde{c}^j(x)) u(x)$$

where the functions $\tilde{c}^j(x)$ are defined by the conditions

$$2(c^j(x) + \tilde{c}^j(x)) = div \, b^j(x).$$

The previous definition is designed to have the following essential properties:

$$\hat{B}^j(uv) = -(\hat{B}^j)^*(uv) \tag{8.1}$$

$$\hat{B}^j(uv) = u\tilde{B}^j v + vB^j u.$$

Equation (8.1) has to be understood in the following sense.

Lemma 8.3 *Let $\hat{B}$ be an operator defined as*

$$\hat{B}u(x) = \sum_{i=1}^{d} b_i(x)\frac{\partial u(x)}{\partial x_i} + \hat{c}(x)u(x).$$

Assume that either $u, v \in H^1(D)$ satisfy $uv = 0$ on ∂D, or that $b(x)$ is tangent to the boundary. Then

$$\hat{B} = -\hat{B}^*$$

(in the sense that

$$\int_D (\hat{B}u)v \, dx = -\int_D u\hat{B}v \, dx$$

for $u, v \in H^1(D)$ specified as above) is equivalent to

$$2\hat{c}(x) = div \, b(x).$$

Proof. We have

$$\int_D (\hat{B}u)v \, dx = \sum_{i=1}^{d} \int_D vb_i \frac{\partial u}{\partial x_i} \, dx + \int_D \hat{c}uv \, dx$$

$$= -\sum_{i=1}^{d} \int_D u \frac{\partial}{\partial x_i}(b_i v) \, dx + \int_D \hat{c}uv \, dx$$

$$+ \int_{\partial D} vu \, b \cdot v \, d\sigma$$

$$= -\int_D uv \, div \, b \, dx + \int_D 2\hat{c}uv \, dx$$

$$- \int_D u\hat{B}v \, dx.$$

Lemma 8.4 *Let* B, $\tilde{B}$, $\hat{B}$ *be operators defined as*

$$Bu(x) = \sum_{i=1}^{d} b_i(x)\frac{\partial u(x)}{\partial x_i} + c(x)u(x)$$

$$\tilde{B}u(x) = \sum_{i=1}^{d} b_i(x)\frac{\partial u(x)}{\partial x_i} + \tilde{c}(x)u(x)$$

$$\hat{B} = \sum_{i=1}^{d} b_i(x)\frac{\partial u(x)}{\partial x_i} + (c(x) + \tilde{c}(x))u(x).$$

Then

$$u\tilde{B}v + vBu = \hat{B}(uv).$$

Moreover, assume that either $u, v \in H^1(D)$ *satisfy* $uv = 0$ *on* ∂D, *or that the vector field* b *is tangent to the boundary. Then*

$$2(c(x) + \tilde{c}(x)) = div\, b(x)$$

implies

$$\hat{B} = -\hat{B}^*$$

(in the sense of the previous lemma).

Proof. *We have*

$$\hat{B}(uv) = \sum_{i=1}^{d} b_i\{\frac{\partial u}{\partial x_i}v + u\frac{\partial v}{\partial x_i}\} + (c + \tilde{c})uv$$

$$= vBu + u\tilde{B}v.$$

The second part of the lemma is just a rewriting of the previous lemma.

Extend all the coefficients a_{ij}, a_i, ..., *to* $\mathbf{R}^d$ *in such a way that they still are of class* C^∞, *satisfy the coercivity condition (5.22), and have compact support. Consider the stochastic parabolic equation in* $\mathbf{R}^d$

$$\begin{cases} d\tilde{u} = \mathcal{A}\tilde{u}\, dt + \sum_{j=1}^{n} \tilde{B}^j\tilde{u}\, dw^j(t) \\ \tilde{u}(0, x) = 1. \end{cases} \tag{8.2}$$

The solution of this equation has P-*a.s. the property* $\tilde{u}$, $\frac{\partial \tilde{u}}{\partial x_i} \in C([0, T]\times\mathbf{R}^d)$, $i = 1, \ldots, d$, *by Theorem 5.9, and* $\tilde{u}(t, x, \omega) \neq 0$ *for all* $(t, x) \in [0, T]\times\mathbf{R}^d$. *The latter fact follows for instance from the representation formula of [37]. Thus, given the domain* $D \subset \mathbf{R}^d$, *there exist two positive random variables* $c_1(\omega) < c_2(\omega)$ *such that,* P-*a.s.,*

$$0 < c_1(\omega) \leq \tilde{u}(t, x, \omega) \leq c_2(\omega) \tag{8.3}$$

for all $(t, x) \in [0, T] \times D$, and

$$|\tilde{u}(t, ., \omega)|_{W^{1,\infty}(D)} \leq c_2(\omega). \tag{8.4}$$

Let $u_0 \in L^2(D)$ be a given initial condition for equation (5.26) or equation (5.46), and let u be the corresponding solution. Recall that these equations can be understood in the following variational sense (cf. [27]):

$$< u(t), \theta >=< u_0, \theta > + \int_0^t a(u(s), \theta) \, ds$$

$$+ \int_0^t < f(u(s)), \theta > \, ds + \sum_{j=1}^n \int_0^t < B^j u(s), \theta > \, dw^j(t) \tag{8.5}$$

for all $\theta \in H_0^1(D) \cap C(\overline{D})$ for equation (5.26), and $\theta \in H^1(D) \cap C(\overline{D})$ for (5.46). Here $< ., . >$ denotes the usual inner product in $L^2(D)$ and $a(u, v)$ is the bilinear form, on $H_0^1(D)$ for (5.26), and on $H^1(D)$ for (5.46), defiend as

$$a(u, v) = \int_D \{ - \sum_{i,j=1}^d a_{ij} \frac{\partial u}{\partial x_i} \frac{\partial v}{\partial x_j} + \sum_{i=1}^d a_i \frac{\partial u}{\partial x_i} v + auv \} \, dx.$$

Moreover,

$$d\phi(x)\tilde{u}(t, x) = \phi(x)\mathcal{A}\tilde{u}(t, x) \, dt + \sum_{j=1}^n \phi(x)\tilde{B}^j \tilde{u}(t, x) \, dw^j(t)$$

for all $\phi \in C^\infty(\mathbf{R}^d)$ (this can be obtained by taking $\theta = \phi\theta'$ in the equation of type (8.5) corresponding to (8.2)). Thus, by Ito formula (cf. [27]),

$$d < u, \phi\tilde{u} >$$

$$=< du, \phi\tilde{u} > + < u, d\phi\tilde{u} >$$

$$+ \sum_{j=1}^n < B^j u, \phi\tilde{B}^j \tilde{u} > dt$$

$$= \{< \mathcal{A}u, \phi\tilde{u} > + < u, \phi\mathcal{A}\tilde{u} > \} \, dt$$

$$+ \sum_{j=1}^n < B^j u, \phi\tilde{B}^j \tilde{u} > dt$$

$$+ \sum_{j=1}^{n} \{< B^j u, \phi \tilde{u} > + < u, \phi \tilde{B}^j \tilde{u} >\} dw^j(t).$$

Now,

$$< \mathcal{A}u, \phi \tilde{u} > + < u, \phi \mathcal{A}\tilde{u} >$$

$$= \int_D \{\phi \tilde{u} \mathcal{A}u + u\phi \mathcal{A}\tilde{u}\} dx$$

$$= \int_D \phi \{\tilde{u} \mathcal{A}u + u \mathcal{A}\tilde{u}\} dx$$

$$= \int_D \phi \{\mathcal{A}(\tilde{u}u) + N(u, \tilde{u})\} dx$$

where

$$N(u, \tilde{u}) = au\tilde{u} - (\nabla \tilde{u})^T \cdot ([a] + [a]^T) \cdot \nabla u.$$

Here [a] denotes the matrix (a_{ij}). We have used the following fact (we shorten the notation for the partial derivatives):

$$\mathcal{A}(fg) = \sum_{i,j=1}^{d} a_{ij} \frac{\partial^2 fg}{\partial x_i \partial x_j} + \sum_{i=1}^{d} a_i \frac{\partial fg}{\partial x_i} + afg$$

$$= \sum_{i,j=1}^{d} a_{ij} \{f_{ij}g + f_j g_i + f_i g_j + fg_{ij}\}$$

$$+ \sum_{i=1}^{d} a_i \{f_i g + fg_i\} + afg$$

$$= g\mathcal{A}f + f\mathcal{A}g - afg + (\nabla g)^T \cdot ([a] + [a]^T) \cdot \nabla f.$$

The previous computation yields

$$d < u, \phi \tilde{u} >$$

$$= < \mathcal{A}(\tilde{u}u) + N(u, \tilde{u}), \phi > dt$$

$$+ \sum_{j=1}^{n} < B^j u \tilde{B}^j \tilde{u}, \phi > dt$$

$$+ \sum_{j=1}^{n} < \hat{B}^j(\tilde{u}u), \phi > \, dw^j(t)$$

recalling the definition of $\hat{B}^j$. This means

$$d(u\tilde{u}) = \{ \mathcal{A}(\tilde{u}u)$$

$$+ N(u, \tilde{u})$$

$$+ \sum_{j=1}^{n} B^j u \tilde{B}^j \tilde{u} \} \, dt$$

$$+ \sum_{j=1}^{n} \hat{B}^j(\tilde{u}u) \, dw^j(t).$$

Therefore, by Ito formula [27],

$$\frac{1}{2} d |u\tilde{u}|^2_{L^2(D)}$$

$$= a(\tilde{u}u, \tilde{u}u) \, dt$$

$$+ < N(u, \tilde{u}), \tilde{u}u > \, dt$$

$$+ \sum_{j=1}^{n} < B^j u \tilde{B}^j \tilde{u}, \tilde{u}u > \, dt$$

$$+ \frac{1}{2} \sum_{j=1}^{n} |\hat{B}^j(\tilde{u}u)|^2 \, dt.$$

The essential fact here is that the Ito term vanishes because of the skew symmetry of $\hat{B}^j$. We have

$$a(\tilde{u}u, \tilde{u}u) + \frac{1}{2} \sum_{j=1}^{n} |\hat{B}^j(\tilde{u}u)|^2$$

$$= - \int_D \sum_{i,j=1}^{d} (a_{ij} - \sum_{k=1}^{n} b_i^k b_j^k) \frac{\partial(\tilde{u}u)}{\partial x_i} \frac{\partial(\tilde{u}u)}{\partial x_j} \, dx$$

$$+ \int_D \tilde{u}u \, N_1(\tilde{u}u) \, dx$$

where N_1 is a first order differential operator,

$$\leq -\eta \int_D |\nabla(\tilde{u}u)|^2_{R^d} \, dx + \frac{\eta}{2} \int_D |\nabla(\tilde{u}u)|^2_{R^d} \, dx$$

$$+C_1 \int_D (\tilde{u}u)^2 \, dx$$

$$\leq -\frac{\eta}{2} \int_D |\nabla(\tilde{u}u)|^2_{R^d} \, dx + C_1 c_2(\omega)^2 |u|^2_{L^2(D)}$$

for some constant $C_1 > 0$. Moreover,

$$< N(u, \tilde{u}), \tilde{u}u >$$

$$+\sum_{j=1}^n < B^j u \tilde{B}^j \tilde{u}, \tilde{u}u >$$

$$\leq C_2 |\tilde{u}|_\infty |\tilde{u}|_{W^{1,\infty}(D)} (|\nabla u|_{L^2(D)} |u|_{L^2(D)} + |u|^2_{L^2(D)})$$

$$\leq \frac{\eta}{8} c_1(\omega)^2 |\nabla u|^2_{L^2(D)}$$

$$+\left[\frac{2C_2 c_2(\omega)^2}{\eta c_1(\omega)^2} + C_2 c_2(\omega)^2 \right] |u|^2_{L^2(D)}$$

for some constant $C_2 > 0$. Note that

$$-\int_D |\nabla(\tilde{u}u)|^2_{\mathbf{R}^d} \, dx$$

$$= -\int_D |\tilde{u}\nabla u + u\nabla\tilde{u}|^2_{\mathbf{R}^d} \, dx$$

$$= -\int_D \tilde{u}^2 |\nabla u|^2_{\mathbf{R}^d} \, dx - \int_D u^2 |\nabla\tilde{u}|^2_{\mathbf{R}^d} \, dx$$

$$-\int_D 2\tilde{u}u\nabla\tilde{u}\nabla u \, dx$$

$$\leq -c_1(\omega)^2 \int_D |\nabla u|^2_{\mathbf{R}^d} \, dx + c_2(\omega)^2 |u|^2_{L^2(D)}$$

$$+2c_2(\omega)^2|\nabla u|_{L^2(D)}|u|_{L^2(D)}$$

$$\leq -\frac{1}{2}c_1(\omega)^2\int_D |\nabla u|^2_{\mathbb{R}^d}\,dx$$

$$+\left[c_2(\omega)^2+\frac{2c_2(\omega)^2}{c_1(\omega)^2}\right]|u|^2_{L^2(D)}.$$

Collecting all these computations, we have

$$\frac{1}{2}\frac{d}{dt}|\tilde{u}u|^2_{L^2(D)}+\frac{\eta}{8}c_1(\omega)^2\int_D |\nabla u|^2_{\mathbb{R}^d}\,dx$$

$$\leq \left[\frac{\eta}{2}\left(c_2(\omega)^2+\frac{2c_2(\omega)^2}{c_1(\omega)^2}\right)+C_1c_2(\omega)^2\right]|u|^2_{L^2(D)}$$

$$+\left[\left(\frac{2C_2c_2(\omega)^2}{\eta c_1(\omega)^2}+C_2c_2(\omega)^2\right)+\lambda c_2(\omega)^2\right]|u|^2_{L^2(D)}$$

$$=c_3(\omega)|u|^2_{L^2(D)}$$

for some positive r.v. $c_3(\omega)$. It follows that

$$\frac{1}{2}\frac{d}{dt}|\tilde{u}u|^2_{L^2(D)}$$

$$\leq c_4(\omega)|\tilde{u}u|^2_{L^2(D)}$$

for some positive r.v. $c_4(\omega)$. Therefore,

$$|\tilde{u}(t)u(t)|^2_{L^2(D)}$$

$$\leq |u_0|^2_{L^2(D)}e^{c_4(\omega)t}$$

which implies that

$$|u(t)|^2_{L^2(D)}$$

$$\leq |u_0|^2_{L^2(D)}e^{c_4(\omega)t}\frac{1}{c_1(\omega)^2}.$$

The proof of the two theorems is complete, recalling lemma 2.2.

Acknowledgements

Most of this work on stochastic flows for infinite dimensional systems has been inspired by Ludwig Arnold and his school. They have developed a large part of the theory and applications of the concept of random dynamical system. I express my gratitude to them, and in particular to Kay-Uwe Schaumlöffel, who shared with me part of his researches on Lyapunov exponents and stochastic flows in infinite dimensions, and to Hans Crauel, who helped me to keep in touch with a variety of motivations for the study of stochastic flows. Moreover, I thank Zdzislaw Brzezniak and Gianmario Tessitore for their collaboration to understand properties of regularity of solutions. I thank my teacher, Giuseppe Da Prato; the semigroup approach to stochastic partial differential equations, used throughout all these notes, is mainly due to him. Finally, I want to thank Jan Seidler, who read these notes in all the details and suggested several improvents and corrections.

References

[1]   L. Arnold (1995) *Random Dynamical Systems*, in preparation.

[2]   L. Arnold, H. Crauel, J.-P. Eckmann, editors (1991) *Lyapunov Exponents*, Proceedings Oberwolfach 1990, vol. 1486 of Lect. Notes in Math., Springer, Berlin.

[3]   L. Arnold, V. Wihstutz, editors (1986) *Lyapunov Exponents*, Proceedings Bremen 1984, vol. 1186 of Lect. Notes in Math., Springer, Berlin.

[4]   Z. Brzezniak (1991) *Stochastic partial differential equations in M type 2 Banach spaces I*, preprint.

[5]   Z. Brzezniak (1991) *Stochastic partial differential equations in Banach spaces II*, preprint.

[6]   Z. Brzezniak, M. Capinski, F. Flandoli (1992) *Stochastic Navier-Stokes equations with multiplicative noise*, Stoch. Anal. Appl. **10**, 523–532.

[7]   Z. Brzezniak, F. Flandoli (1992) *Regularity of solutions and random evolution operator for stochastic parabolic equations*, in: Stochastic Partial Differential Equations and Applications, G. Da Prato and L. Tubaro eds, Research Notes in Math. 268, Pitman, Harlow, 54–71.

[8]   Z. Brzezniak, F. Flandoli (1994) *Almost sure approximation of Wong-Zakai type for stochastic partial differential equations*, to appear on Stoch. Processes and their Appl.

[9]   H. Crauel, F. Flandoli (1994) *Attractors for random dynamical systems* Prob. Theory and Related Fields **100**, 365–393.

[10]   G. Da Prato (1983) *Some results on linear stochastic evolution equations in Hilbert spaces by the semigroup method* Stoch. Anal. Appl. **1**, 57–88.

[11]   G. Da Prato (1986) *Equations aux derivees partielles stochastiques et applications*, Ecole Polytechnique, Centre de Math. Appl., R.I. 148.

[12]   G. Da Prato, M. Iannelli, L. Tubaro, (1982) *Some results on linear stochastic differential equations in Hilbert spaces* Stochastics **6**, 105–116.

[13] G. Da Prato, L. Tubaro, (1985) *Some results on semilinear stochastic differential equations in Hilbert spaces* Stochastics **15**, 271–281.

[14] G. Da Prato, J. Zabczyk (1992) *Stochastic Equations in Infinite Dimensions*, Cambridge Univ. Press, Cambridge.

[15] M.C.Delfour, S.K.Mitter (1972) *Hereditary differential systems with constant delays 1. General case*, J. Diff. Eq. **12**, 213–235.

[16] F. Flandoli (1990) *Dirichlet boundary value problem for stochastic parabolic equations: compatibility relations and regularity of solutions*, Stochastics **29**, 331–357.

[17] F. Flandoli (1991) *Stochastic flows and Lyapunov exponents for abstract stochastic PDEs of parabolic type*, in: Lyapunov Exponents, L. Arnold, H. Crauel, J.-P. Eckmann eds, LNM 1486, Springer-Verlag, 196–205.

[18] F. Flandoli (1990) *Solution and control of a bilinear stochastic delay equation*, SIAM J. Control Optimiz. **28**, 936–949.

[19] F. Flandoli (1992) *On the semigroup approach to stochastic evolution equations*, Stochastic Anal. Appl. **10**, 181–203.

[20] F. Flandoli (1994) *Stochastic flows for nonlinear second order parabolic SPDE*, preprint Scuola Normale Superiore, n. 35.

[21] F. Flandoli, K.-U. Schaumlffel (1990) *Stochastic parabolic equations in bounded domains: random evolution operators and Lyapunov exponents*, Stochastics **29**, 461–485.

[22] A. Ichikawa (1982) *Stability of semilinear stochastic evolution equations*, J. Math. Anal Appl. **90**, 12–44.

[23] N. Ikeda, S. Watanabe (1981) *Stochastic Differential Equations and Diffusion Processes*, North Holland, Amsterdam.

[24] P. Kotelenez (1982) *A submartingale type inequality with application to stochastic evolution equations*, Stochastics **8**, 139–151.

[25] N.V. Krylov (1992) W_2^n – *theory of the Dirichlet problem for SPDE in a bounded domain*, preprint.

[26] N.V. Krylov, B.L. Rozovskii (1981) *Stochastic evolution equations*, J. Sov. Math. **16**, 1233–1277.

[27] Kunita (1990) *Stochastic Flows and Stochastic Differential Equations*, Cambridge Univ. Press, Cambridge.

[28] O.A. Ladyzenskaya, V.A. Solonnikov, N.N. Uraltzeva (1967) *Equations de Types Paraboliques Lineaires et Quasi-lineaires*, Moskow.

[29] I. Lasiecka (1980) *Unified theory for abstract parabolic boundary problems- A semigroup approach*, Appl. Math. Optim. **6**, 287–333.

[30] J.L. Lions, E. Magenes (1972) *Non-Homogeneous Boundary Value Problems and Applications*, Springer-Verlag, New York.

[31] S. Mohammed (1984) *Stochastic Functional Differential Equations*, Research Notes in Math. 99, Pitman, London.

[32]   E. Pardoux, *Equations aux Derivée Partielles Stochastiques nonlineaires monotones*, These, Université Paris XI, 1975.

[33]   E. Pardoux (1982) *Equations du filtrage non lineaire, de la prediction et du lissage*, Stochastics **6**, 193–231.

[34]   A. Pazy (1983) *Semigroups of Linear Operators and Applications to Partial Differential Equations*, Springer-Verlag, New York.

[35]   B.L. Rozovski, A. Shimizu (1981) *Smoothness of solutions of stochastic evolution equations and the existence of a filtering transition density* Nagoya Math. J. **84**, 195–208.

[36]   D. Ruelle (1982) *Characteristic exponents and invariant manifolds in Hilbert space* Ann. Math. **115**, 243–290.

[37]   K.-U. Schaumlffel, F. Flandoli, (1991) *A multiplicative ergodic theorem with applications to a first order stochastic hyperbolic equation in a bounded domaun*, Stochastics **34**, 241-255.

[38]   A.V. Skorohod (1984) *Random Linear Operators* Reidel, Dordrecht.

[39]   D.W. Stroock, S.R.S. Varadhan (1979) *Multidimensional Diffusion Processes* Springer-Verlag, New York.

[40]   H. Tanabe (1979) *Equation of Evolution*, Pitman, London.

[41]   G.M. Tessitore (1994) *Existence, uniqueness and space regularity of adapted solutions to SPDE*, preprint Scuola Normale Superiore n. 28.

[42]   P. Thieullen (1987) *Exposants de Lyapounov des fibres dynamiques pseudo-compact* Ann. Inst. Henri Poincare', Analyse Nonlineaire **4**, 49–97.

[43]   H. Triebel (1978) *Interpolation Theory, Function Spaces, Differential Operators*, North-Holland, Amsterdam.

[44]   L. Tubaro (1988) *Some results on stochastic partial differential equations by the stochastic characteristic method* Stoch. Anal. Appl. **6**, 217–230.

INDEX

adjoint equation 41, 47, 63
adjoint flow 62
adjoint operator 43, 59
analytic semigroup 12, 13, 24, 31,
43, 49
approximation 11, 14, 60, 61
bilinear equation 19
boundary value problem 2, 11, 23,
41, 42, 44, 48, 50, 51, 66–68
bounded domain 1, 2, 7, 18, 19,
22, 31, 42, 43
cocycle property 58
coercivity 8, 20, 29, 32, 33,
38–41, 44–48, 70
compatibility relation 18, 51, 52, 55
continuous version 3, 4
deterministic boundary value problem 51
differential operator 2, 10, 18, 22, 26,
29–36, 39, 40, 42, 44, 46, 69, 74
Dirichlet boundary condition 19, 22, 31,
41–44, 50–52, 66, 67
elliptic problem 49, 50
ellipticity 26, 29, 31, 35, 41, 43, 44,
66, 67
evolution property 58
Feynman-Kac formula 7
fractional power 24
geometric restriction 7, 18, 68
Green mapping 49
Hilbert scale 17, 24, 48, 49,
Hilbert-Schmidt modification 57
Hilbert-Schmidt operator 5, 13, 57, 58, 63
Hilbert-Schmidt stochastic flow 57, 58, 63
infinitesimal generator 10, 11, 13, 49
interpolation space 16
linear random field 5

Lyapunov exponents 2, 58, 76
mild form 24, 57
mild solution 19, 24,
Neumann boundary condition 33, 43,
52, 67
non-homogeneous boundary value
problem 10, 48
optimal regularity result 53,
parabolic equation 1, 2, 11, 12,
18, 20, 22, 25, 29, 31, 50, 51,
66, 68, 70,
pathwise Green formula 59,
random field 3–5, 57
random fundamental solution 7
regular version 2, 3, 5, 6, 57
regularity of flow 2
regularity theory 18, 24, 27, 31, 66
robust equation 7
scale of Hilbert spaces 8
semilinear equation 7
skew-symmetry 7
stochastic characteristics 8
stochastic delay equation 6,
stochastic evolution equation
1, 6, 9
stochastic flow 1–3, 5–8, 20,
57, 58, 63, 65
strongly continuous semigroup
10, 11
transposition 1, 2, 62, 66
well posed 24–30, 33, 37–41,
44–48, 58, 59, 63, 65
well posed and coercive 25, 27
41, 42, 44, 45, 48
Wiener integral 5, 6
Yosida approximation 14, 60

Printed in the United States
by Baker & Taylor Publisher Services

Printed in the United States
by Baker & Taylor Publisher Services